W9-AOM-788

INTRODUCTION TO

PHYSICS

Amanda Kent and Alan Ward

Consultant: Dr M. P. Hollins

Designed by Iain Ashman

Edited by Jenny Tyler

Computer program by John Freeman

Illustrated by Sue Stitt, Chris Lyon, Jeremy Gower, Simon Roulstone and Mick Gillah.

Contents

Additional illustrations by Jim Bamber, Jeremy Banks, Hayward & Martin, Rob McCaig, Elaine Mills, Martin Newton, Graham Round, Graham Smith and Ian Stephen. Additional designs by Glenn Lord and Stanley Sweet.

First published in 1983 by Usborne Publishing Ltd, 20 Garrick Street, London WC2E 9BJ, England.

Printed by
Graficas Reunidas SA,
Madrid, Spain.

What is physics?

Physics is to do with all the things around you and the energy they have. It is about why things get hot, what light is, how things move and make sounds, and so on.

It was the Ancient Greeks who first studied science and some of the ideas of physics come from them. Even the word

"physics" comes from an Ancient Greek word. Many of the basic laws and principles of physics are several hundred years old, but this doesn't mean they are old-fashioned or out of date. Most modern scientific discoveries are based on them and you need to know about them to understand how anything from a bicycle to a spaceship works.

The main areas of physics are light, heat, sound, mechanics, electricity and magnetism and this book has sections on each of these. There are experiments to help you understand some of the important ideas in physics. These are designed so that

you can find most of the things you need at home or in a nearby shop. If you find that an experiment doesn't work first time, don't worry – this often happens in science. The conditions may not be quite right and are interfering without you realizing. Just try again.

While reading this book, try and think about things around you and see how they fit in with the ideas you are reading about. You may think of your own experiments to do, too, to test the things you read about.

Towards the back of the book there is a physics computer program written to work on most common makes of home computer. If you have one of these or can borrow one from someone, type it in and try it out. It is all about using electricity in the home.

At the very back of the book are pages of physics words, where you can find definitions of some of the words you will meet in the book and the proper wording of some of the laws, such as Newton's Laws.

You will find answers to most of the questions and puzzles at the back of the book too. Some questions do not have answers; they are things for you to think about.

All about energy

The world you live in is full of energy: light, heat, electricity and sound are some of the forms that energy takes. You use your own energy to move about and do your work. Most energy comes from the Sun, which provides heat and light for plants to grow, to keep you warm and let you see. Even fuels, such as oil and gas, were made from plants that absorbed the Sun's energy as they grew, millions of years ago.

Potential and kinetic energy

Food you eat and petrol in a motorbike are forms of stored energy that can be used to make you or the motorbike move. These are both "potential energy" and they change to "kinetic energy" when things move.

Chemical energy

Fuel in rockets and explosives in fireworks have potential chemical energy, changing to kinetic energy when rockets take off and fireworks explode.

Gravitational energy

If you put something high up, you give it potential energy, which has come from your muscles. If it falls, its potential energy changes to kinetic energy.

Strain energy

All solids, but especially springs and elastic, have this. The energy is potential when something is stretched or squeezed and changes to kinetic energy when let go.

Chemical energy

Gravitational energy

Strain energy

Scare your friends

Find a piece of stiff card that fits into a long envelope which opens at one end. Cut a square in the card and loop a rubber band over the card. Put a small piece of card between the band and stick a piece of paper to each side of the card trapping the rubber band.

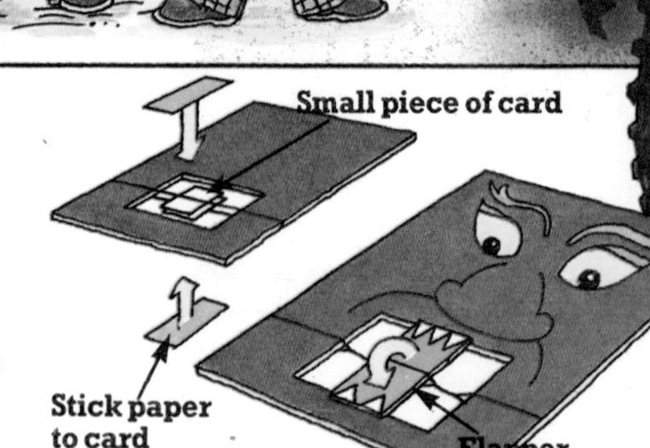

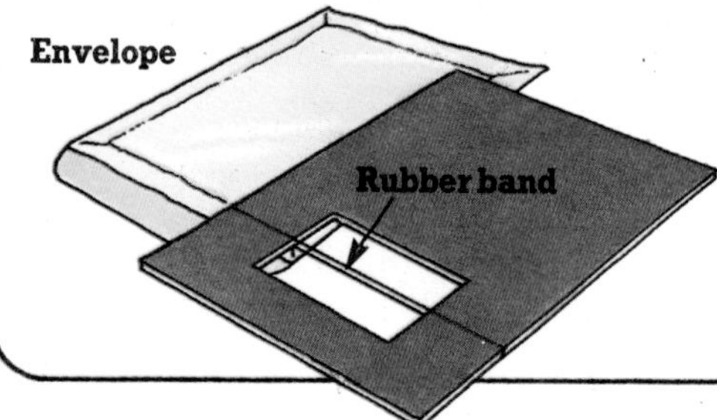

Wind the "flapper" many times to give it potential energy. Put the card into the envelope and give it to your friends. When they pull the card out, potential energy changes to kinetic as the flapper starts to spin round.

Sound energy goes into phone

Energy is never lost or made. It just changes from one kind to another. A telephone takes in sound energy. This changes to electrical energy for the journey and is then changed back to sound energy again.

Electrical energy changes back to sound energy at other end.

Phone changes this to electrical energy.

Solar panels use Sun's energy

Windmill uses wind energy.

People have developed many ways of getting the sorts of energy they need from other sorts of energy, e.g. using wind and Sun energy for their homes. This is what is meant by "harnessing" energy.

Measuring energy

Energy is measured in units called joules (J for short). These units are named after a man called J. P. Joule, of Manchester, who found out that heat was a form of energy. For the man in the picture to lift the heavy weight above his head, he uses about 1,000J of chemical energy from his body.

Energy puzzle

The dog is at the top of the stairs (1). He then runs down (2) to his dinner, and eats it (3). See if you can work out the energy changes that are happening in this picture and what kind of energy the dog has at each stage. (Answer on page 47).

Light energy

Nearly all the energy you need comes from the Sun; it is a source of light and heat energy. There are other sources of light, such as light bulbs, but most of the things you see do not give off light. Light coming from a source bounces off them, and some of that light goes into your eyes and makes you see them.

Do you know which of these things are sources of light? Turn to page 47 to find out if you were right.

SUN MIRROR
GOLD GLASS
TORCH SILVER
CANDLE
MOON FOIL

1 How does light move?

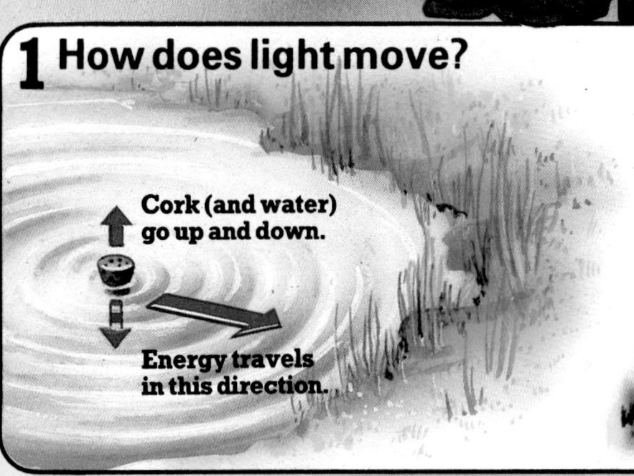

If you try to see how light is actually travelling you will find it impossible. Physicists believe it travels in some ways like water, as a wave motion. They think light energy is carried along very tiny ripples, much smaller than water waves.

Imagine a cork on a pond. Waves make the cork bob up and down but it doesn't move in the direction the wave is going. Light waves are the same; each part is vibrating up and down like the cork while the light energy moves along the wave.

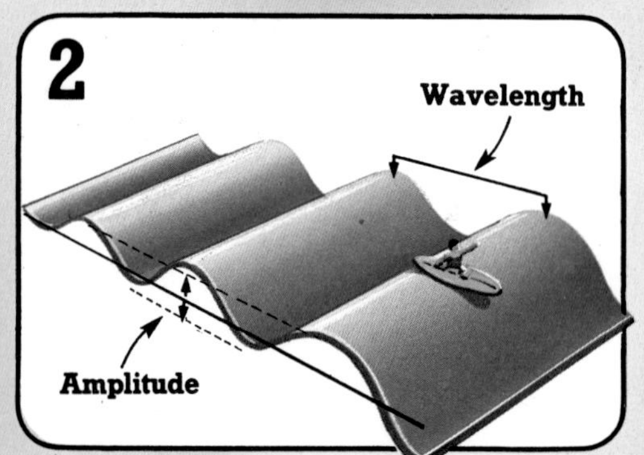

In order to compare waves, three measurements are taken: wavelength, which is the length from the top of one wave to the top of the next; amplitude, which is the height of the wave, and frequency, which is the number of waves that pass a certain point each second.

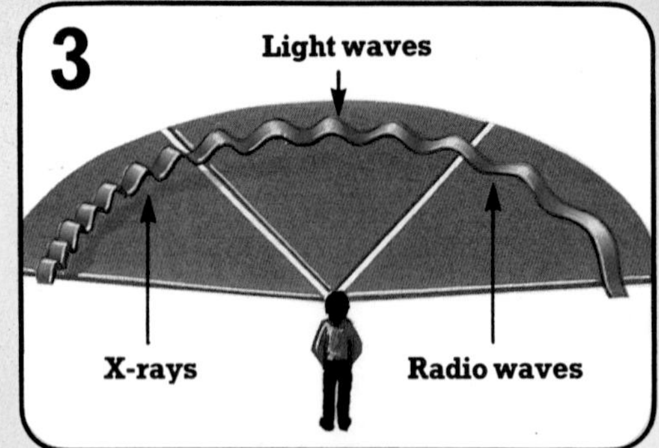

Light waves belong to a family of waves called the electromagnetic spectrum*. This includes X-rays, television, radio and heat waves. They all travel at the same speed, but their wavelengths are different and they have different effects on things.

**See page 40 for more about the electromagnetic spectrum.*

Shadows

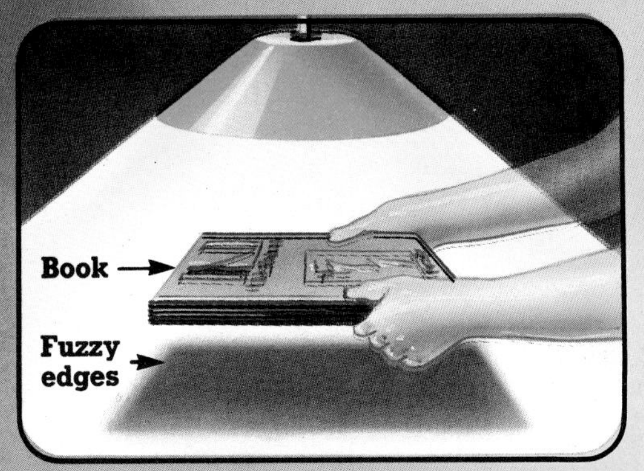

Some things, such as glass and air, allow light to pass straight through them, and are called transparent. When light hits something opaque (which means that the light waves are not able to pass through) a shadow is made where the light does not reach.

Hold a book under a light. You will see a shadow beneath it which has slightly fuzzy edges. Light waves hitting the book are bounced off. The shadow has fuzzy edges because the light bulb is large. Each point on the light bulb gives a sharp shadow, but in a slightly different place.

On a sunny day, see how long your shadow is at midday, and then late in the afternoon. The length of your shadow depends on the angle at which the light is hitting you. Try to think of light as travelling as masses of tiny waves from a source. These waves move in straight lines until they hit something. Then they make a shadow where the light cannot reach.

Shadows can be useful

You can make a sundial by cutting out a circle of white card, pushing a knitting needle through the centre and sticking the needle into the ground. Mark where the Sun's shadow falls each hour, and you have made a "clock".

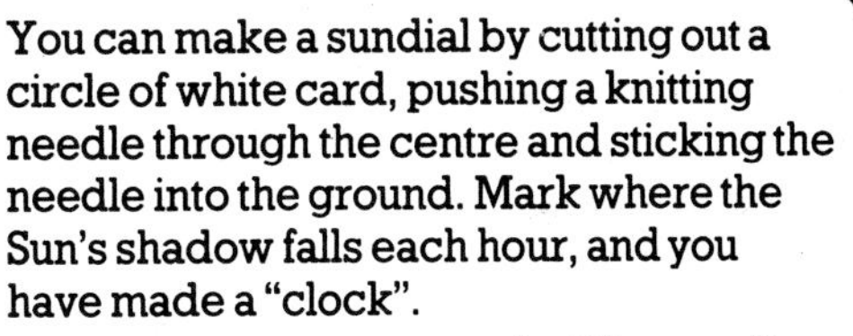

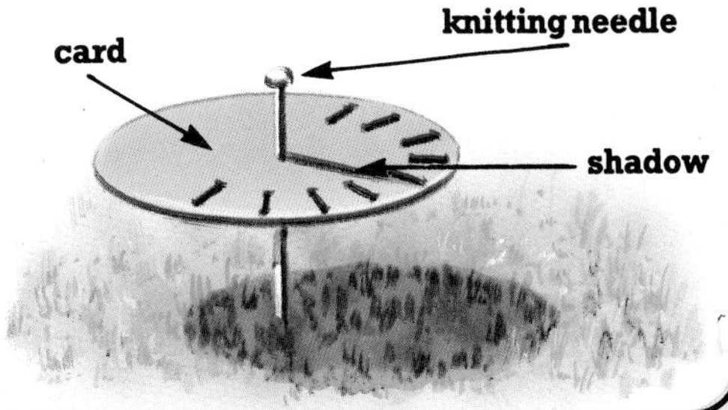

Shadows used to help people tell the time before clocks were invented. They used sundials. Some people still have them in their gardens. The time is "read" from the sundial, either by the length or the position of the shadow. It only works, of course, when the Sun is shining.

How fast does light travel?

Light travels at an enormous speed. It covers 300 thousand kilometres in every second, which is more than 42 thousand times as fast as Concorde.

Seeing things

Eyes and cameras work in the same way – light travels into them and makes images inside. If you make this "pinhole camera" you will see how this works.

1 Find an old shoebox, make a hole 4 cm across in one end, cover it with black paper, sticking it down inside.

2 Cut away most of the other end and cover this with greaseproof paper to make a screen. Put the lid on and stick it down with tape.

3 Make a pinhole through the black paper and put the box near to a light bulb. Look at the screen to see an upside down image of the light bulb.

Screen

Light rays

Pinhole

What is happening?

Light travels in straight lines from the light bulb. Some of this light goes through the pin hole. Light from the top of the light bulb will hit the bottom part of the screen, and light from the bottom of the light bulb will hit the upper part of the screen.

The bigger the hole, the fuzzier the image.

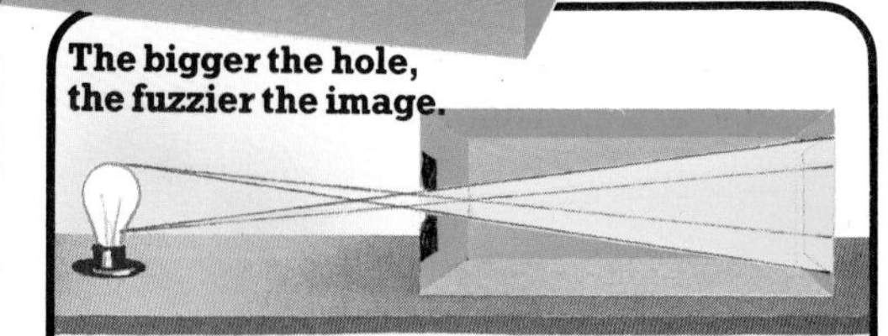

Make the hole a bit bigger and the image will be much more fuzzy. This happens because light coming from each part of the light bulb is able to make an image on quite a wide area of the screen, so the different images fall on top of each other and cause a fuzziness.

If you make two other pinholes, you will find there are two more images on the screen. Because only a small amount of light can pass through each hole, you get clear, separate images on the screen.

These lines are called light rays. They show where the light is going.

Taking a photograph with the pinhole camera

To take a photograph, remove the greaseproof paper and find a well fitting back for the box. Under red light put some photographic paper in the back of the camera, and stick the camera back down well. Cover the pinhole with your finger.

Place the box so that the hole is facing the light bulb. Let the light fall on the photographic paper by taking your finger away for about one minute, then put the box under red light again.

Remove the paper, move it around in a dish of "developer" until the picture appears. Dip it in water to wash it, put it into a dish of "fixer", and wash it again for 20 minutes.

What lenses do

The photographs that can be taken with a pinhole camera are not very clear. This is because there is no lens in it. There are lenses in your eyes, in magnifying glasses, telescopes, microscopes and cameras. In all these, the lens bends the light so the light rays can all meet at a particular point. Lenses are always transparent, and they have curved edges.

Try on the glasses that some of your friends wear. You will discover that some are much stronger than others; the stronger glasses will make your eyes feel odder.

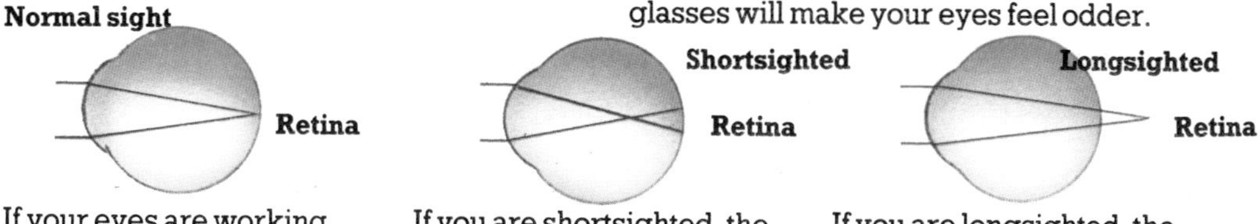

If your eyes are working properly, the light rays coming from a point meet at a point on the retina, which is at the back of your eye.

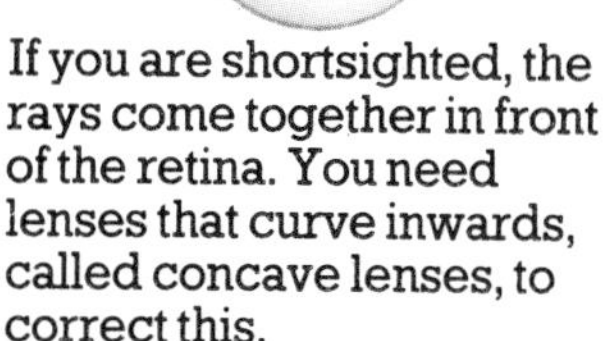

If you are shortsighted, the rays come together in front of the retina. You need lenses that curve inwards, called concave lenses, to correct this.

If you are longsighted, the light rays come together behind the retina. You need glasses with lenses that curve outwards, called convex lenses, to correct this.

Eyes and cameras make images in the same way as a pinhole camera.

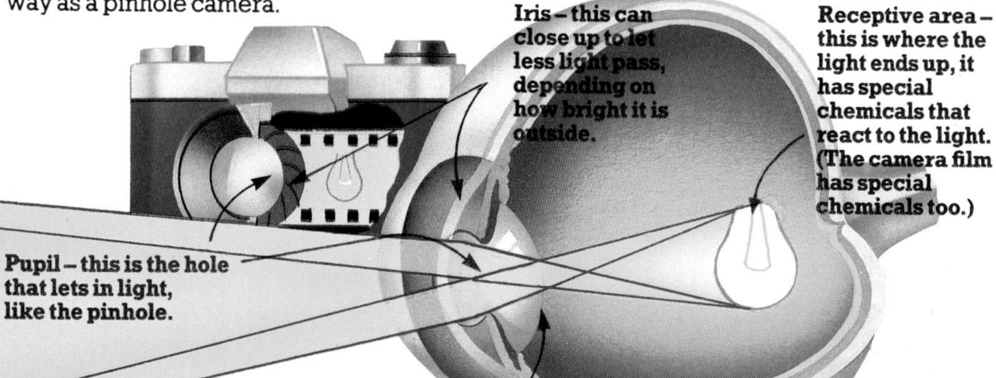

Reflection

Lots of things around you reflect light: windows, glasses, cars that have been well polished, shiny boots, a pool of still water, silver foil. But reflection is best in mirrors, because they are so smooth and shiny.

Stand next to a friend in front of a mirror. Can you see how they look different from how you normally see them? This is because the mirror changes everything around. Try winking your right eye, and it seems that your left eye is winking back at you in the mirror. What you see in the mirror is a "wrong-way-round" image of yourself.

The law of reflection

If you throw a ball straight at a wall, it will bounce straight back. If you throw it at an angle it will bounce off at the same angle. Try it yourself and see. This is the law of reflection, and it works for light too.

Draw two lines at equal angles to a flat mirror. Shine a ray of light (you could use a torch) along one of these lines. You will see that the mirror reflects the ray along the other line. The ray and its reflection will always make the same angle with the mirror.

Angle of incidence

A line at right angles to the mirror at the point where the ray strikes is called the normal.

The angles between the rays and the normal are called the angles of incidence and reflection.

Angle of reflection

The law of reflection can be written: angle of incidence = angle of reflection.

Put something in front of a small mirror, such as some dice. Move the mirror away and notice how the image of the dice moves the same distance away in the mirror. This is always true: the image is the same distance from the mirror as the object.

Image distance

Object distance

Refraction

Light waves can travel through transparent things, but they slow down when they enter them. It is rather like when you walk into the sea; the water slows you down. Light travels fastest in air, 25% slower in water and 35% slower in glass.

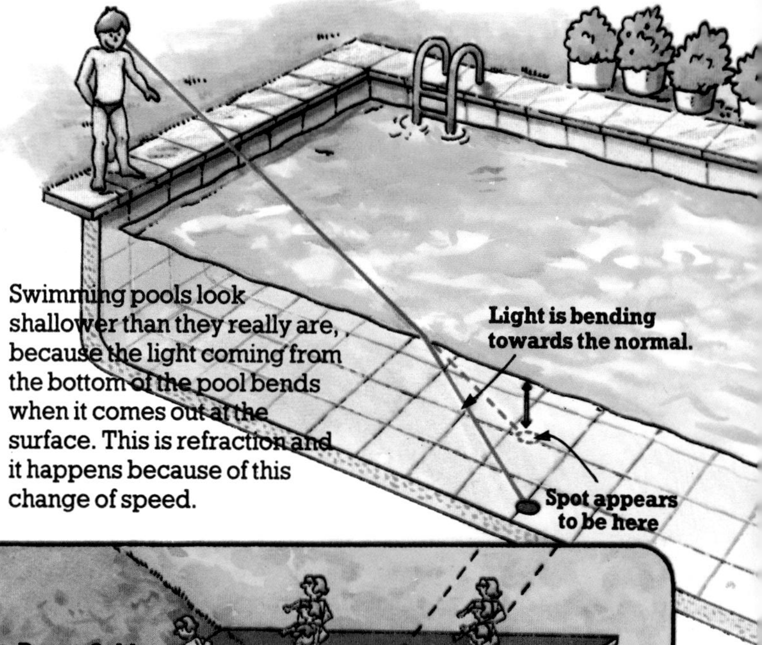

Swimming pools look shallower than they really are, because the light coming from the bottom of the pool bends when it comes out at the surface. This is refraction and it happens because of this change of speed.

Light is bending towards the normal.

Spot appears to be here

Why light bends

The first soldiers out of the mud go faster. So the direction changes again.

Boggy field

Soldiers go slower.

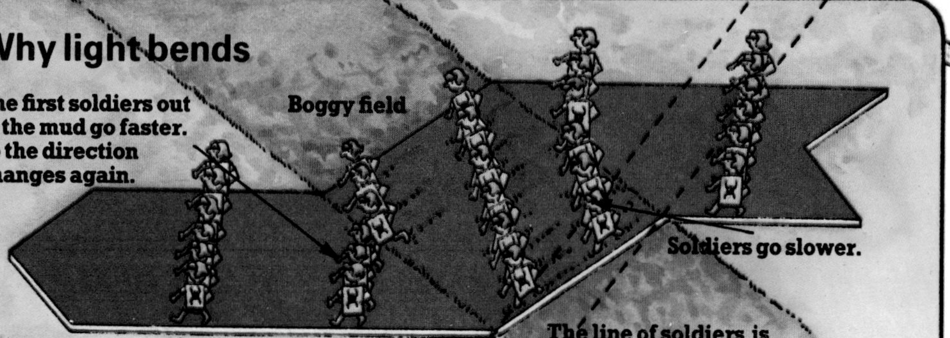

The line of soldiers is like light entering a swimming pool.

When a straight line of soldiers march into a boggy field their direction will change. The first soldiers to march into the mud will slow down before the others. When rays of light hit the surface of glass or water at an angle, the first ones to enter will be slowed down before the others. This makes light bend as it enters glass or water. It also bends, in the opposite direction, when it leaves glass or water.

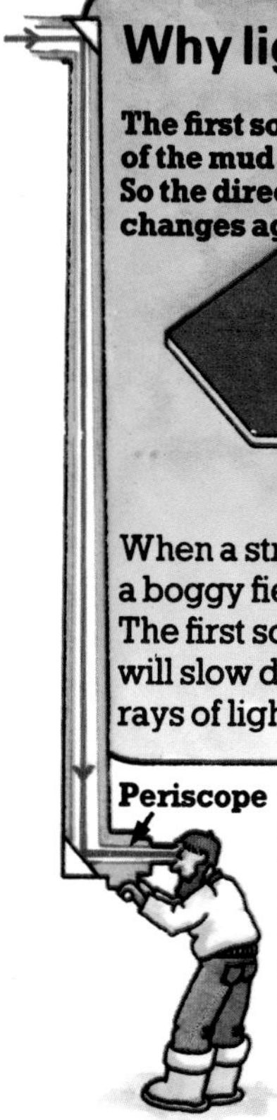

Periscope

Total internal reflection makes periscopes work and also fibre optic cables, which are hair-thin strands of glass. Light travels along the fibres by being totally internally reflected from side to side.

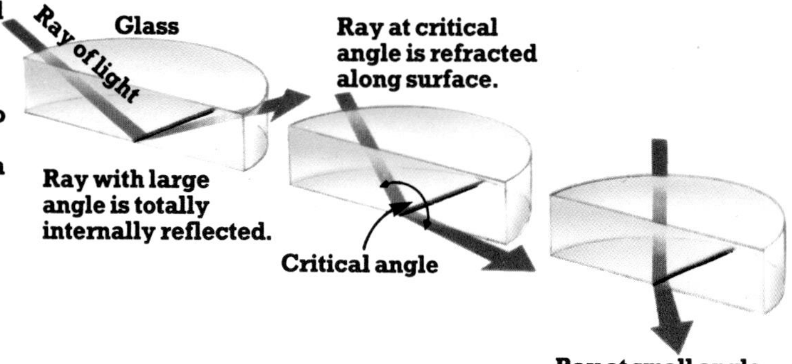

Ray of light

Glass

Ray with large angle is totally internally reflected.

Ray at critical angle is refracted along surface.

Critical angle

Ray at small angle is refracted.

Sometimes, light can't get out of water or glass because it is hitting the surface at a very big angle. The light bounces back into the water or glass. This is called "total internal reflection" and it can be very useful.

Light sometimes comes out of the glass or water, but travels along the surface. It only does this when it hits the surface at a particular angle called the "critical angle". This angle is different for glass and water.

Colour

Ordinary white light is a small group of waves in the electromagnetic spectrum.* It is a mixture of different colours, and each colour has a different wavelength**

Isaac Newton discovered that light is made up of different colours in 1666. He let a beam of sunlight coming through a hole in his window blind pass through a glass prism. (This is a triangular chunk of glass.) The prism split the light up into a band of colours on his wall. He called the band of colours the "solar spectrum".

Make a spectrum

When rays of sunlight pass through raindrops they split into different colours too. The raindrops are acting as prisms. On a bright sunny day you could try making a spectrum on the wall. (Either early morning or late afternoon is best, when the sun is low in the sky.) Position a mirror inside a plastic box filled with water. Put it opposite a window facing the Sun. The wedge of water between the surface and the mirror acts as a prism.

If your wall is not white, tape or hold a sheet of paper over the place where the colours will appear.

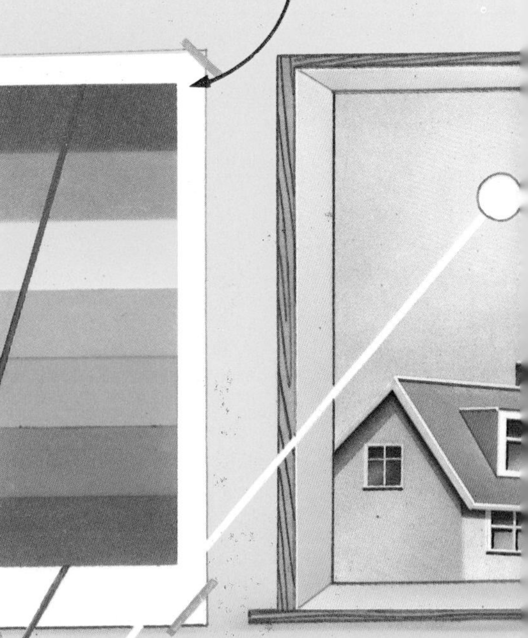

The wedge of water bends each different wavelength by a slightly different amount. Red has the longest wavelength and is bent the least. Violet has the shortest and is bent the most. The colours always appear in the same order: red, orange, yellow, green, blue, indigo, violet. (Remember, ROY-G-BIV.)

Juggle the position of the box and the slant of the mirror, until you get a solar spectrum on the wall.

You can see how the colours mix to make white light by waggling your fingers in the water. The colours blur and become white.

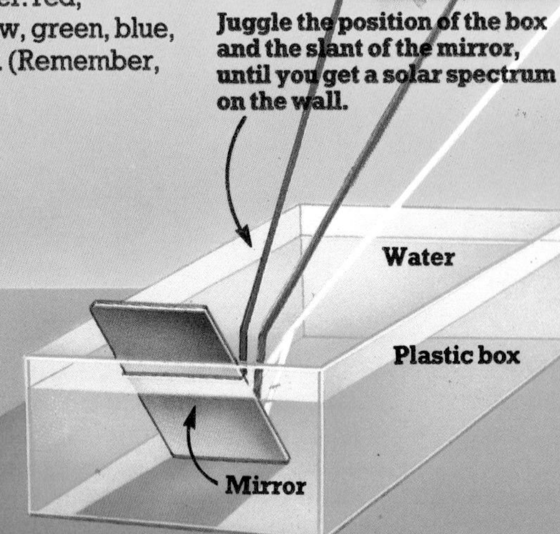

Although we say the spectrum is made up of the colours red, orange, yellow, green, blue, indigo and violet, each of these really consists of a whole range of wavelengths. The yellow band is lots of different yellows, for example, ranging from orangey yellows to greeny yellows.

*See page 40 for more about this.

**See page 6 for more about wavelengths.

Colour mixing

You can mix colours in two ways: either by mixing different coloured lights together or by mixing up paints.

The three main coloured lights are red, green and blue. These are the three primary colours in science. If red, green and blue lights are mixed together they make white light, as is shown on the right.

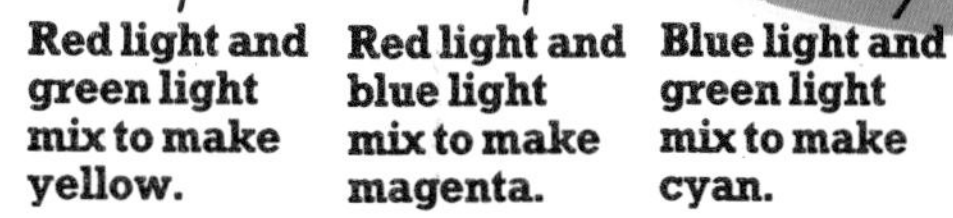

Colour television pictures are made up from these three primary colours. The picture consists of millions of tiny glowing dots: some red, some blue and some green. The light from the glowing dots mixes to form all the different colours you see on the screen.

Paints and all coloured things contain pigments. When we say something is coloured red, it is because the pigments are absorbing all the colours in the light that is hitting it except the red, which is reflected. A blue object has pigments that absorb all the colours in white light except blue.

Red light

Why leaves are green

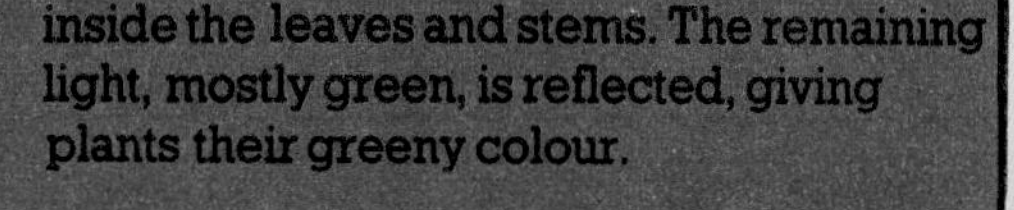

The chemical processes that go on in plants need mainly red light. As long as a plant is alive, the red light from sunlight is absorbed by a pigment called chlorophyll inside the leaves and stems. The remaining light, mostly green, is reflected, giving plants their greeny colour.

Make a colour mixer

Cut out a circle of card, about 8cm across. Draw seven equal-sized segments on it and colour them in the colours of the rainbow. Make a hole in the middle and stick a pencil through it, point downwards. Try spinning your top. What colour is it when it is spinning fast? Why do you think this is?

card

pencil

(Don't worry if it looks a bit dirty. This is because your colours are not very pure.)

Heat energy

Heat is another form of energy and is also measured in joules. It travels as masses of tiny waves, in the same way as light does, at the same speed, but with a different wavelength. It can't be made, but comes from other sorts of energy: from electricity, for example, in an electric heater. Heat is often a by-product of other energy changes, such as when a gun is fired. In this case heat and sound energy are given off as by-products.

Here you can see what happens when water changes its "state", that is, changes from solid to liquid to gas.

What does heat do?

Absolutely everything around you is made of particles called atoms which are much too tiny to see. They are usually joined into little groups called molecules. These vibrate backwards, forwards and sideways all the time with kinetic energy. They even vibrate in solids, though not enough to lose their places in a neat pattern. When heat waves hit the molecules, the energy of the waves changes to kinetic energy which makes the molecules move around even more. The molecules knock against each other, rather like marbles hitting one another, and the vibrating energy is passed on from one to another.

See how molecules move

Put some peas in a jar and shake the jar very gently. The peas will vibrate, but will stay in roughly the same places. This is what happens when a solid is heated. Shaking the jar more will give the peas more energy. They will roll over each other like the molecules in a liquid. If you shake harder some of the peas may even jump out of the jar. This is what happens when a liquid boils: some of the molecules jump out and make a gas.

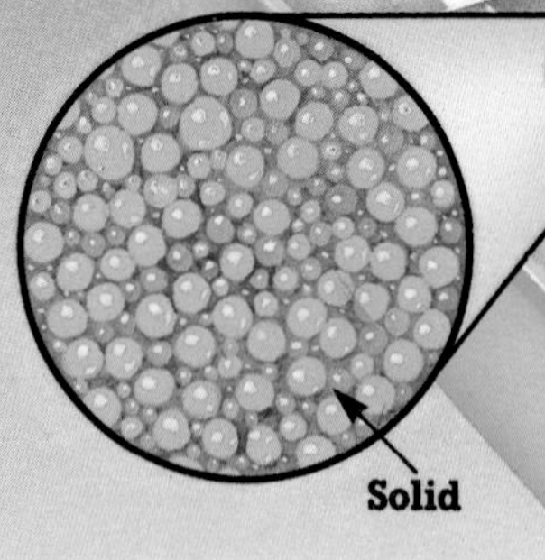

1 SOLID

The molecules in ice are vibrating a tiny bit. It is only when they are warmed that they get enough energy to move around and become water.

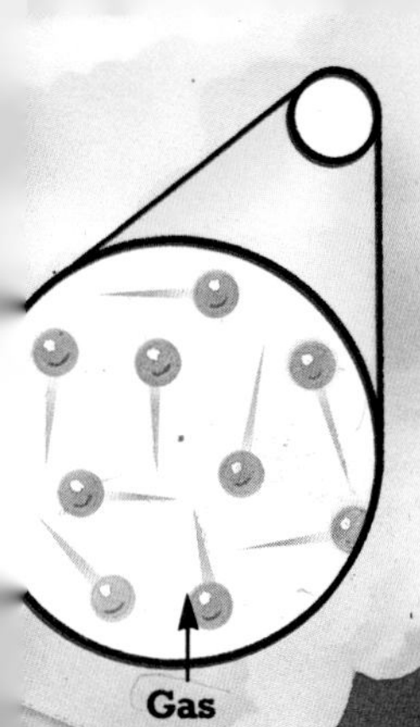

3 GAS
Steam takes up more space than water, which is why the lid of the saucepan rattles about. Steam, which is gas, contains molecules that are flying around in the air. If they hit something cold they will change back to water again. They pass their energy to the cooler surface, which, as a result, is warmed up slightly.

2 LIQUID
As water gets hotter and hotter, the molecules have more and more energy and are able to move further and faster. Some soon have enough energy to jump out of the liquid. When the water is boiling, lots of the molecules have enough kinetic energy to leave the water as steam.

Liquid

Why do you think a saucepan boils over sometimes?

Water is odd

In most substances, the liquid takes up more space than the solid because the molecules move further apart. Water is odd because when a lump of ice melts, the water that is left occupies *less* space. This happens because of the way the molecules are arranged in ice. Pipes sometimes burst in winter because freezing water expands and cracks them.

Can you tell how hot something is?

Your body is not very good at this. Fill three bowls with water, one with hot, one with cold and the other with warm water. Put one hand in the hot water, and the other in cold water for a few seconds. Then put them both in the warm water and your hands tell you different things. The one that was in hot water will think the warm water is very cold and the one that was in cold water will find it hot.

Temperature is the measurement of how hot something is and, as we cannot measure it ourselves, we need something to do it for us. Thermometers, such as the special medical one below, are used for this.

Narrow part – this gives you time to read the thermometer because once the mercury has passed this point it can't go down again until the thermometer is shaken.

Bulb – full of liquid mercury. When the surroundings get hotter, its molecules move around more and the liquid expands up the tube.

Scale of temperature – this shows the temperature of the surroundings measured in degrees Celsius. 0°C is the temperature of ice and 100°C the temperature of boiling water. Our body temperature does not vary much from 37°C so this thermometer only goes from 35°C to 42°C.

There are several other kinds of thermometers. Some use a special sort of alcohol to measure very low temperatures, others work by using gas. Temperature can even be measured using electricity.

How does heat travel?

Heat energy can travel in waves in the same way as light. This is called heat radiation. Heat waves travel from the Sun at 300 million metres (about eight times round the Earth) per second, to reach us across about 240 million kilometres of empty space. The journey takes about eight minutes. The more "red hot" something gets the more heat it radiates. Electric fires, hot plates and light bulbs all radiate heat.

The heat waves themselves are not hot, but when they are stopped and absorbed by something, that thing will get hot. Dark-coloured things absorb more radiant heat that light things. If a swimming pool is solar heated, black panels are used. The water from the pool circulates under the panels and gets warm because the panels absorb the Sun's heat well.

Radiant heat is reflected by white and shiny surfaces. People tend to wear pale colours in summer because the heat is reflected off them. In hot countries, such as Australia, many of the cars are white. Try touching light and dark coloured cars when they have been in the hot sunshine. The dark ones are much hotter.

Rising heat

As liquids and gases get warmer, their molecules have more kinetic energy and so can move further apart. They are now less "dense"* and each bit is lighter, so they can rise up. Cold liquid or gas, being more dense, sinks down. When heat is carried by liquids and gases in this way it is called convection.

Gliders use convection currents.

How radiators work

Central heating radiators give off most of their heat by convection, not radiation. They warm the air around them which rises as convection currents. The cold air sinks and is warmed in turn.

**See page 25 to find out more about density.*

Birds soar upwards on convection currents.

When land is warmer than the air above, air warmed by the land rises as convection currents. Gliders and birds use convection currents to keep them up in the air. Birds can soar upward without flapping their wings at all when they are in a convection current.

Cool air sinks down.

Hot air rises.

Houses need to be well ventilated and air must circulate round the rooms. As air is warmed by heaters and fires, it becomes less dense, rises towards the ceiling, where it is mixed with cool air coming in the windows. It then sinks again.

Invisible heat movement

Heat can actually move through some things without you knowing it is happening. It does this by movements of molecules. When molecules are heated they have more kinetic energy which they pass on by bumping against each other. When heat moves this way it is called conduction.

Some things are better at conducting heat than others. Air is a very bad conductor, and so are most clothes. In cold weather people wear woollen sweaters. The warmth of their bodies can only escape very slowly. In hot countries people wear very loose cotton clothes. The heat outside is not conducted towards their bodies because of the air between their skin and their clothes. Convection currents inside the clothes take the hot air away.

Spoon test

Which spoon do you think will get hottest? The butter melts quickest on the silver spoon. Silver is the best conductor of heat here. The plastic spoon will get least hot. It is a bad conductor of heat, which is why saucepans often have plastic handles. Substances that are poor conductors of heat are called insulators.

Puzzle

How is your home kept warm? Do you have radiators, open fires, gas fires, double glazing? Are you kept warm by radiation, conduction or convection? Or by all three?

Sounds and noises

Sound is another form of energy. All sounds happen because something is vibrating, which makes molecules in the air begin to vibrate too. The molecules themselves are not the sound, but without them there is silence.

What happens when you make a sound?

Try this with your ruler . . .

Bend up and let go . . .

Book

Ruler

1. When the ruler is up, air molecules are squashed together above the ruler and thinned out underneath.

2. When the ruler is down, molecules are crowded together underneath and thinned out above.

3. Meanwhile, the first group of close-together molecules is now expanding and pushing the next group of molecules above it together. The vibrating ruler is pushing the air into a pattern of molecules that are at first close together, then far apart.

How you hear sound

You hear a humming noise because the pattern of molecules moving through the air hits your eardum and makes it vibrate too. These vibrations trigger off tiny pulses of electricity which travel along nerves to your brain. Your brain interprets these pulses as sound.

Sounds can only travel if there are molecules around. In space there are no molecules so astronauts have to talk to each other by radio. (Radio waves, like light waves, can travel where there are no molecules.)

Amplitude

Frequency

The more waves on the screen, the higher the frequency, and the higher the pitch of sound. Things that vibrate fast have a higher pitch.

What sound . . .

Scientists use an oscilloscope, which looks like a small television, to see the pattern that sound makes. Sound vibrations are changed to electrical vibrations inside a

How fast does sound travel?

Sound travels much faster through solids and liquids than it does through air. You can tell a train is about to arrive because the rails hiss before you hear the train.

Did you know that American Indians put their ears to the ground when they are listening for horses? Why do you think this is?

Underwater sounds

By sending out sound pulses, ships can detect whether something such as a submarine is below them in the water and how deep it is. The sound pulses are reflected back to the ship when they hit something. This is called sonar. Sound travels four times as fast in water as in air.

Resonance

If you "ping" a glass with your finger, it will vibrate and ring. It does so with its own special frequency, called its natural frequency. A singer, singing a note of the same frequency, is supposed to be able to make a glass vibrate so much that it cracks. This is called resonance.

Noises

Noises, such as heavy traffic, are made up of a jumble of vibrations of different frequencies. Their vibrations do not follow a regular pattern as those of other sounds do. The loudness of sounds and noises can be measured in decibels (dB). Very high decibel counts can damage people's ears and may cause deafness. Here you can see how loud some everyday sounds are.

...looks like

microphone and these are used to make wave shapes appear on the screen. The crests show where a big group of molecules is hitting the microphone.

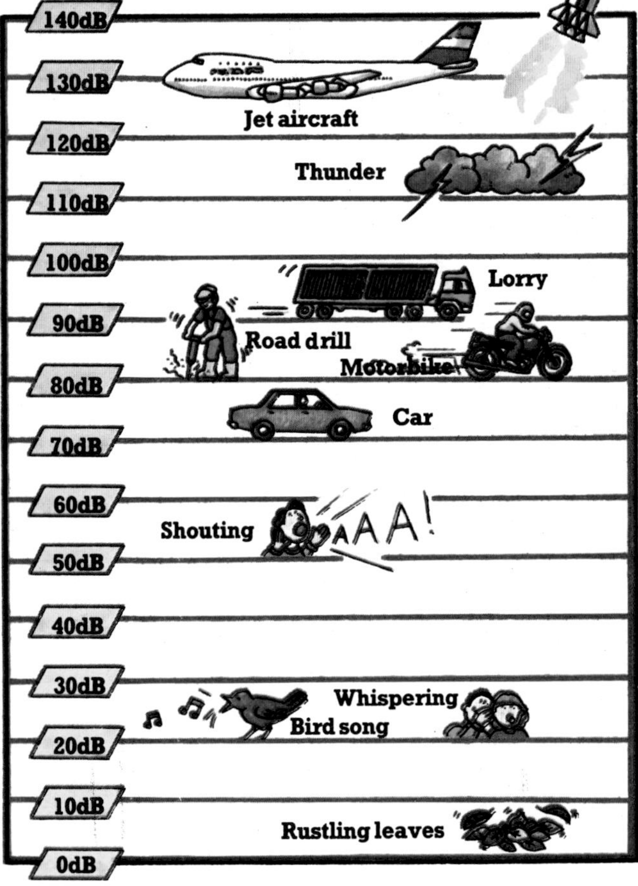

Music

There are three main types of musical instrument, and in each one the notes that can be produced depend on what is vibrating.

Blowing

All wind instruments work by making a column of air vibrate. The pitch of the notes can be varied by changing the length of the air column. Try blowing across the tops of bottles containing water.

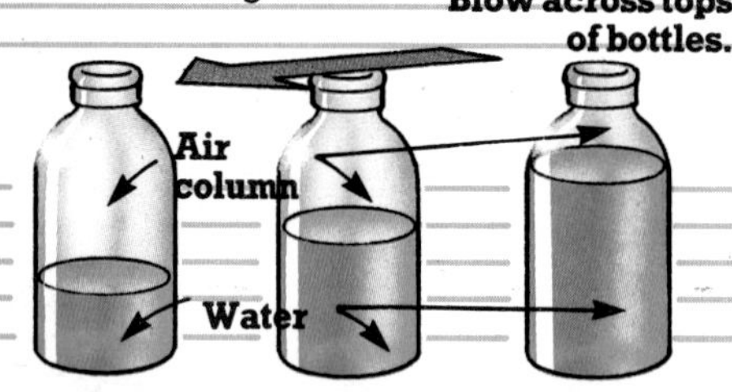

Plucking strings

When the strings of a guitar are plucked they vibrate and the air around them starts to vibrate too. If you put your fingers on the string you shorten the length of the bit that can vibrate, and this makes the pitch higher. Making strings tighter or using lighter strings also raises the pitch. Try stretching an elastic band around a book and two pencils. Change the vibrating length of the elastic band by moving your finger along its length.

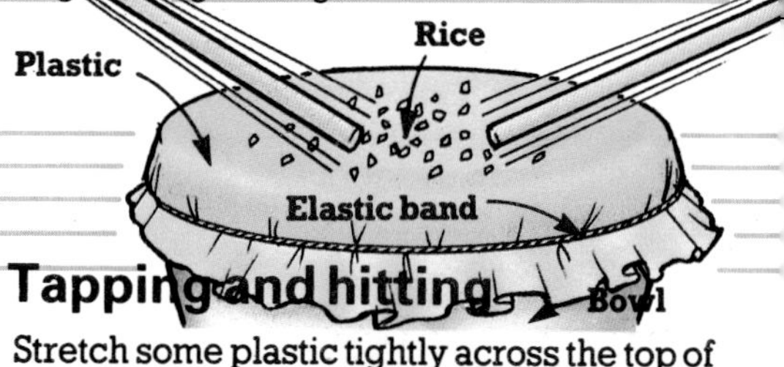

Tapping and hitting

Stretch some plastic tightly across the top of a bowl. Put some sugar or rice on the plastic, tap the top and see how the grains vibrate. Drums make "sounds" because their covering vibrates, sending sound waves into the air.

Music puzzle

Can you think how all these things produce music – by blowing, plucking strings, or hitting? (Answers on page 47.)

How music is stored

Cassette tapes store sound as a code made up from arrangements of iron oxide particles. To do this, music is played near a microphone, which sets up electric pulses. These pulses magnetize the iron oxide particles into a code representing the sound of the music.

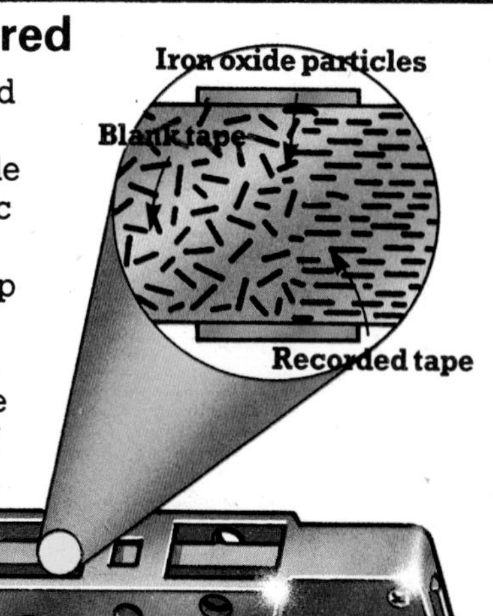

To make a record, the message on a "master tape" is converted back to electric pulses, which are fed to a "cutting head". This has a very sharp diamond point. It vibrates with the electric pulses and cuts grooves into a soft plastic disc. Loud sounds make the groove deeper and high notes make it more wavy. This plastic disc is used as a sort of mould. The records you buy are made from imprints of it.

Electric music

When the strings of an electric guitar are plucked, their vibrations are converted to electrical energy and sent to an amplifier. Here the electrical signals are magnified and then sent to a loudspeaker where they are changed into sounds.

Synthesizers

Synthesizers make music by using electric signals instead of vibrations. They are usually connected to a keyboard. Each note played on the keyboard sends a particular electric message to the synthesizer. This then sets up an electrical code for the sound that is wanted. The code is sent to an amplifier and then a loudspeaker, which converts it to vibrations in the air that you can hear.

Computer music

Several microcomputers have a tiny synthesizer inside the keyboard, which enables them to play simple tunes and make sounds. You have to type in a command such as "sound" or "ping" followed by the note you want and how long you want it played for.

Instructions from the micro

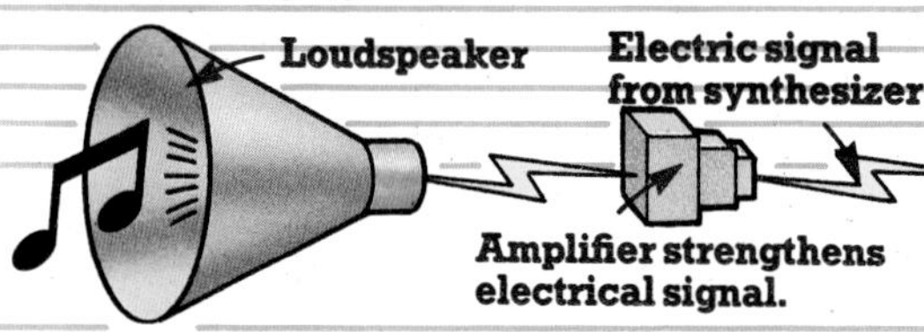

Synthesizer

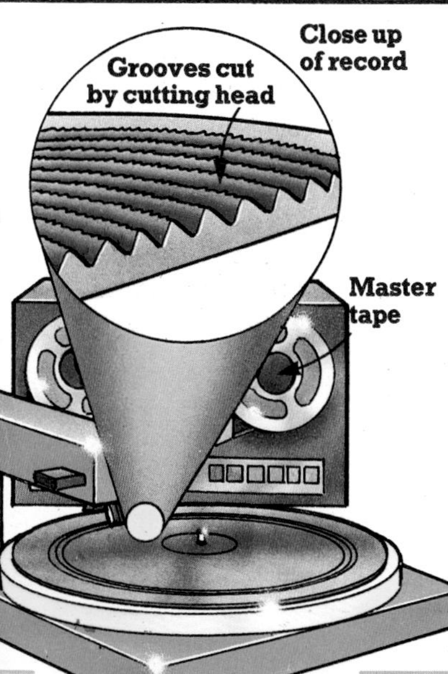

Compact discs

A very modern way of storing sound is on compact disc. These are only 120mm wide (a normal long playing record is 300mm wide). Instead of grooves there are millions of microscopic "pits and flats". Turntables for the discs have a laser beam instead of a stylus. The beam scans the surface of the disc, "reading" the pattern of pits and flats and converting this into electrical signals and then vibrations.

Laser beam

Pits

Flats

Close up of compact disc

Mechanics

Mechanics is not just about garages. It is about everything that is happening to things – how heavy they are, what is pushing and pulling them, how they move and what they can do. The next ten pages are all about mechanics.

Forces

We are always pulling, pushing and lifting. A push or a pull is called a force. A force can get an object moving, or stop it from moving, or change the direction of its movement, or squeeze it and change its shape.

Force is measured in newtons (N), after a very famous scientist, Sir Isaac Newton (1642-1727). A force of one newton is quite a small force. The picture on the left shows people exerting forces and gives a rough idea of the size of these forces.

Gravity

Newton's name is connected with the study of the force of gravitation – the pull we usually call gravity. He began wondering about this pull when an apple fell on his head. There is a force between all objects that pulls them together. Usually this force is small, but because the Earth is so big it has a strong pull. It pulls things like the apple towards it by the force of gravity.

The Moon is much smaller than Earth, and its gravitational pull is only one sixth of the pull of Earth's gravity. On the Moon you could kick a ball six times as far, and jump six times as high.

Your weight = your mass x pull of the Earth.

How heavy are you?

Kilograms measure your mass, the "amount of stuff" inside you, and this is the same wherever you are. When you stand on the scales, your mass is being checked against "standard" masses. There is a standard kept of every single measurement somewhere in the world, which means people will always know exactly what these measurements are.

Weight is a measure of the pull of gravity on you, and is measured in newtons because it is a force. To convert your mass into your weight you multiply it by the pull of the Earth (about 10 units) and give your answer in newtons. So, if your mass is 60kg, your weight is about 600N.

How much would you weigh on the Moon? And what would your mass be? Don't forget, the pull of the Moon is only 1/6 that of the Earth. (Answer on page 47).

Centre of gravity

The Earth's gravity pulls down on every single particle with a force equal to the weight of that particle. In a body, the forces seem to be concentrated at the "centre of gravity". Once the centre of gravity is hanging outside the base of the object, it will fall over.

Boat falls over because its centre of gravity has moved "outside" its base.

You put your arms out when you are trying to balance. By holding them out and moving them up and down, you are able to alter your centre of gravity so that it is still over your base (your feet) and you don't fall over. Why do you think a tight rope walker uses a pole?

Stability

It is difficult to make something fall over if it is stable. Stable things have very low centres of gravity. Racing cars are built very low on the ground with a low centre of gravity so they can keep upright when cornering fast. Perhaps you can think of other stable things.

An empty plastic bottle is not very stable. You can knock it over easily because its centre of gravity is high. If some water is added, the weight in the bottom of the bottle makes the centre of gravity lower and it is now more stable. If the bottle is full of water, the centre of gravity is higher and it is not very stable again.

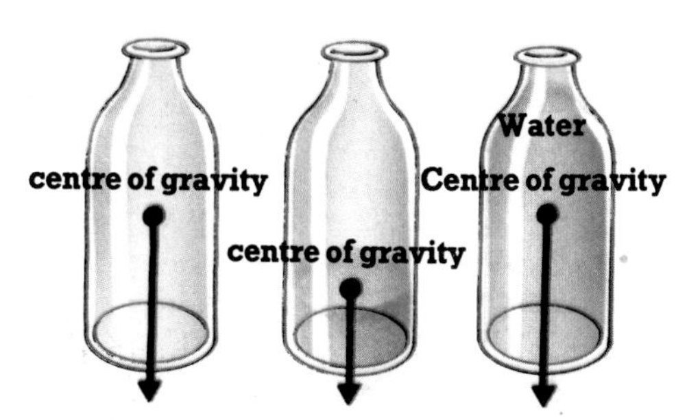

What is pressure?

Pressure is how much push there is on a certain area. Atmospheric pressure, for example, is measured by working out the weight of the air in newtons pressing down on a square metre of ground. It is given in newtons per square metre (N/m^2) and at sea level atmospheric pressure is $100,000N/m^2$.

Anything, solid, liquid or gas, can exert pressure. The pull of gravity makes you push on the ground – through the area of your shoes that are touching the ground's surface. When a doctor measures your blood pressure, he measures how much the blood is being forced along your arteries and veins. Try

pushing your thumb into a piece of wood. You probably won't make a mark, but if you pushed with the same force on the top of a drawing-pin, you would be able to push deep into the wood. Your push is now concentrated over the much tinier area of the point.

Peel a banana

Put about 1cm of methylated spirit into a bottle and light it by dropping in a match. Put the peeled end of the banana in the mouth of the bottle. At first the heated air expands, forcing air out of the bottle. When the flame goes out, air inside the bottle cools and contracts, and its pressure is reduced. Then the greater force of atmospheric pressure outside the bottle pushes the banana inside, peeling it at the same time.

Banana

Meths (only use a tiny bit)

Liquids can push too

Water and all liquids take the shape of whatever container they are in. All the time they are pressing outwards on the inside of the container, trying to escape.

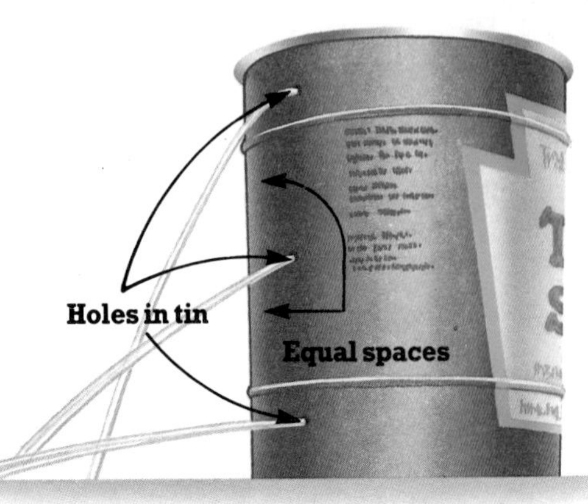

Pressure experiment

Make three holes the same distance apart up the side of a tall tin, cover the holes with a strip of sticky tape and fill the tin with water. Stand it so its holes are next to a sink and pull off the tape. Water spurts out further from the lower holes because it is pressed out with a stronger push. This pressure is caused by the weight of water that lies above. So the deeper the water, the greater the pressure. (Try a taller tin and check this for yourself).

Water molecules attracting each other.

Pond skater

Surface tension

Molecules in liquids attract each other and pull towards each other in all directions. The molecules at the surface, though, having nothing above them to attract, so they pull together much more on each side. This makes the surface layer act rather like a light skin. This is called surface tension. This skin is really quite strong. Some insects, such as the pond skater, can run across the surface of water.

See surface tension happen

Arrange some matches on water in a bowl. If the surface in the centre is touched with blotting paper the matches move towards the centre. Blotting paper soaks up some of the water and the whole surface of the water including the matches is pulled towards the centre. If you touch the centre with a piece of soap the matches move away.

What is happening?

Some of the soap molecules dissolve in the water in the centre and the soap and water molecules mix. The pulling forces between water molecules break down, which means the surface tension is reduced and the water is pulled back from the centre by stronger forces on the edges.

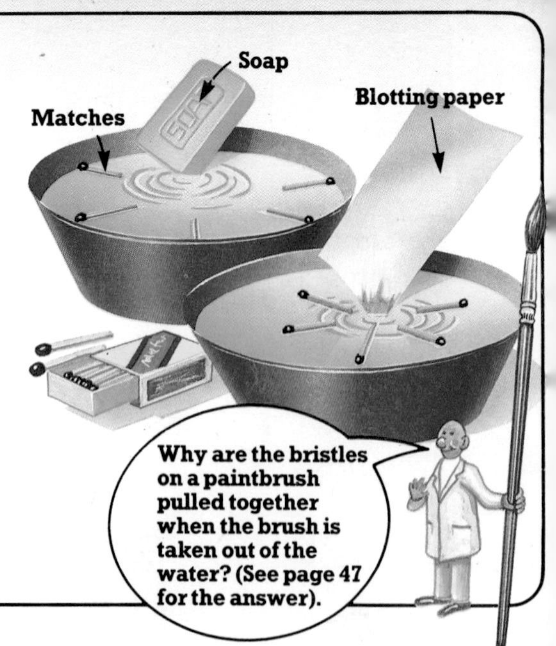

Why do we use soap?

Water molecules are more strongly attracted to each other than to other things. When soap is added the special "wetting agent" it contains breaks down the surface tension of the water molecules, which means they can wet things better.

Bubbles

Bubbles are like elastic envelopes made from layers of soap or washing-up liquid and water. The air inside is slightly compressed and is pushing out from the centre equally in all directions. The liquid has two surfaces, both of which are pulling inwards in all directions.

Why do things float?

Whether something floats depends on how heavy it is compared to its volume. This is called its density. Scientists say that the density of water is 1, and then compare the densities of everything else to it. Only if its "relative density" is less than 1, can something float in water. Here are some relative densities.

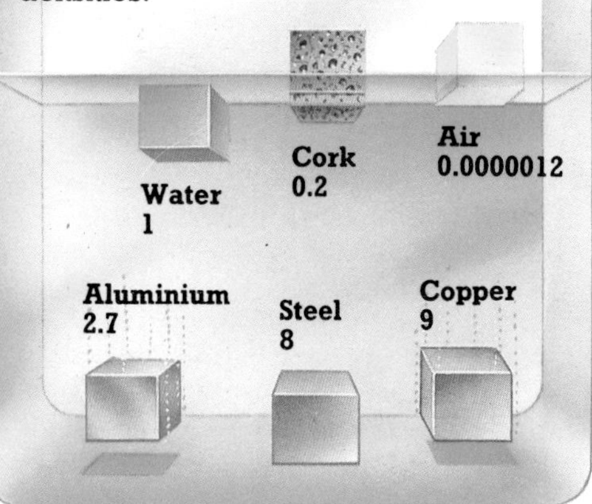

How does a steel ship float?

Steel has a much higher relative density than water, yet a steel ship can float. Look at the plan of a ship. It is not a solid lump of steel. There are lots of empty rooms full of air. Its *average* density is less than water. People can float in water, so our *average* density must be less than water too.

Plan of ship

Displacement

When things float, the weight of water they push to the side (displace) is equal to their weight. You float better and displace less water when you breathe in. This is because having air in your lungs makes your average density less. Why do you think submarines take water into special tanks when they want to dive deep under the sea?

The weight of the ship is the same as the weight of the water it has pushed aside.

Stopping and going

Over 300 years ago Isaac Newton worked out a set of rules that explain the way in which things move. These rules can apply to anything, even the most modern machinery. You can see below that the way a pair of roller skates moves is to do with Newton's laws. As you read through these two pages, try and think of other things, like cars and trains, and how they move.

Starting off

1. To make anything start moving, go faster or stop, a force is needed. This person needs a force (a push from his friend) to start him sliding across the ice on a tray.

Push (force) needed to start moving

2. Once moving, the person would carry on for ever at the same speed in a straight line unless a force acted upon him. This is Newton's first law*. Friction is important here – friction is a force that happens when two things rub together, such as the tray rubbing against the ice. Friction opposes motion and makes things slow down.

Once moving a force is needed to stop things. (Friction is often the force that does this.)

3. The person accelerates slowly up to a certain speed and also takes a little time to decelerate again. The time taken to change speed (accelerate or decelerate) depends on the size of the person. A bigger person would take longer. This resistance to change of movement is called "inertia". The bigger the person the more inertia he has.

Quicker and faster

Newton found that things speed up (accelerate) quicker if the force pushing is greater. The person will accelerate faster across the ice if his friend pushes harder. If the person had less mass, the same force would accelerate him further. This is Newton's second law* of motion.

Forwards and backwards, up and down

Whenever there is a force on one thing in one direction, another force is acting on something else in the opposite direction. This is Newton's third law.* When a gun is fired, at the same time that the bullet shoots out of the barrel, the gun kicks back into your shoulder. Somebody pushing a person across the ice on a tray will find themselves falling backwards as the person on the tray sets off forwards.

*See page 46 for the exact wording of Newton's laws of motion.

Friction can be helpful

When you skate on ice, your skates move easily. There is very little friction between your skates and the ice because the ice is smooth and your skates are sharp. But on roads you need to grip the surface in order to walk. You need friction. Roads are not smooth and shoes and tyres are made with ridges on them to make more friction between them and the road.

Friction in liquids

There is friction between the layers of molecules in some liquids such as treacle, honey and oil, which is why they are sticky and slow to pour. They are called viscous liquids; and some of them, such as oil, can be very useful for stopping two pieces of metal in a machine rubbing together. Oil is put between the metals so that instead of lots of friction between the pieces of metal there is only the small amount of friction in the oil.

Why do you think we don't use water to reduce friction?

Liquid inertia

Liquids have inertia too. You can use this fact to tell the difference between a hardboiled egg and a raw egg. Spin them both on separate saucers. Stop them with your fingers, then let go almost immediately. The raw egg starts spinning again. This is because the layers of liquid inside the raw egg are still spinning round with "moving inertia".

Roller skates

Here you can see how Newton's laws of motion affect roller skating.

The skater's muscles produce the force that he needs to push against the air resistance, to go uphill or to accelerate. Once he is moving, if there were no forces acting such as air resistance and friction, he would carry on for ever (Newton's first law).*

The more the person pushes against the ground, the faster he will accelerate (Newton's 2nd law).*

Bigger people take longer to build up speed.

Gravel thrown backwards as skates go forwards (Newton's 3rd law).*

Greasing wheels reduces friction.

Speed, acceleration and gravity

Speed is how fast you are going. It is the distance you travel in a certain time. Velocity is different from speed. It is a measure of how fast you are going in a particular direction. Both speed and velocity are measured in metres per second (m/s) or more commonly kilometres per hour (km/h). When you change velocity you either go faster (accelerate), slower (decelerate) or change direction. Acceleration and deceleration are both measured in metres per second per second (m/s^2), because you are changing velocity with time.

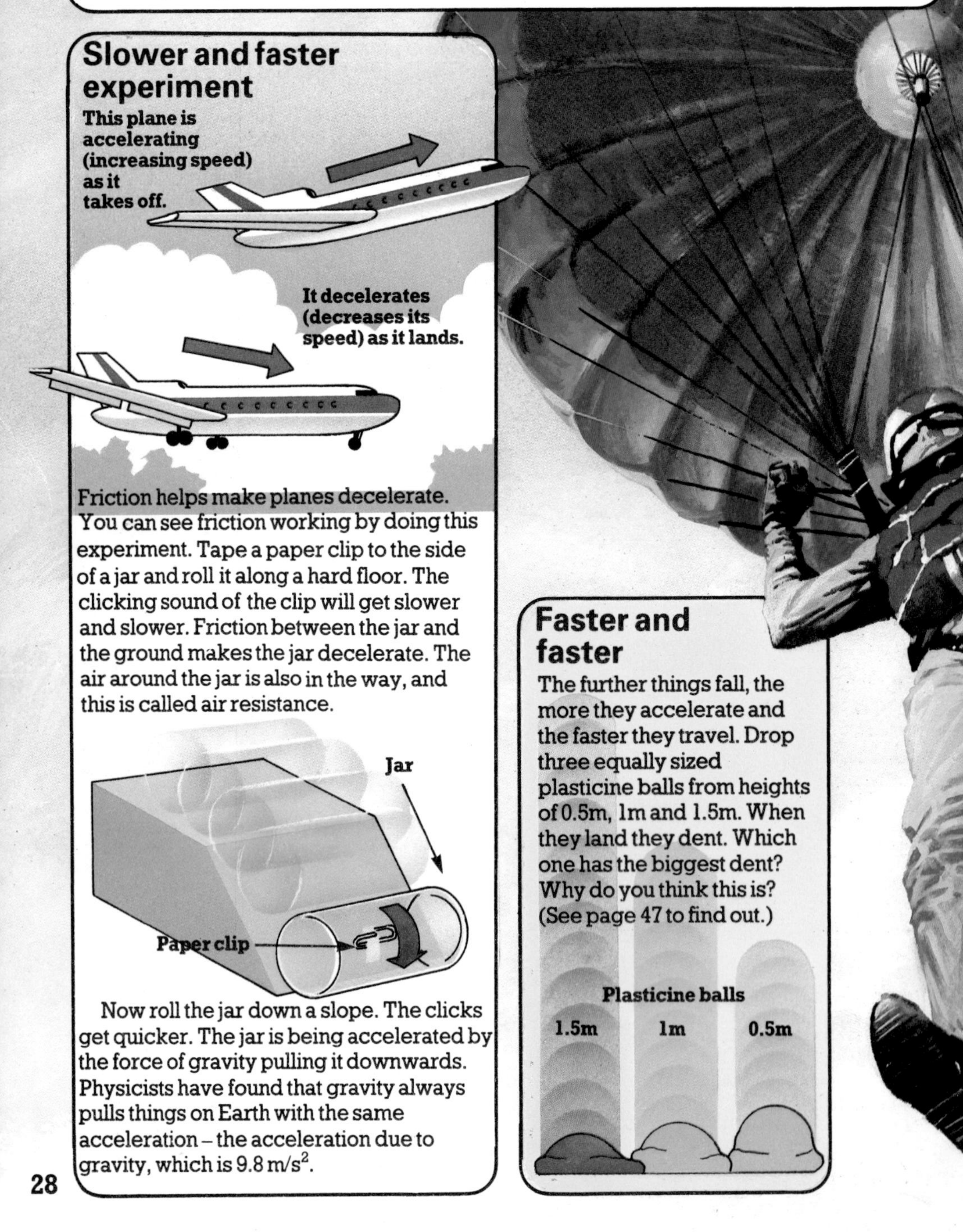

Slower and faster experiment

Friction helps make planes decelerate. You can see friction working by doing this experiment. Tape a paper clip to the side of a jar and roll it along a hard floor. The clicking sound of the clip will get slower and slower. Friction between the jar and the ground makes the jar decelerate. The air around the jar is also in the way, and this is called air resistance.

Now roll the jar down a slope. The clicks get quicker. The jar is being accelerated by the force of gravity pulling it downwards. Physicists have found that gravity always pulls things on Earth with the same acceleration – the acceleration due to gravity, which is $9.8 m/s^2$.

Faster and faster

The further things fall, the more they accelerate and the faster they travel. Drop three equally sized plasticine balls from heights of 0.5m, 1m and 1.5m. When they land they dent. Which one has the biggest dent? Why do you think this is? (See page 47 to find out.)

The Earth and the Moon

The Moon (like all satellites) circles around the Earth, keeping the same distance away and travelling at the same speed. It doesn't need a force to push it round as in space there is no friction at all, so once it is going at a certain speed it will continue for ever. It is kept the same distance from Earth all the time because of the Earth's strong gravitational pull.

Earth

Moon

This distance never changes.

Centrifugal force

Centrifugal force comes from the Latin words meaning "centre" and "flee". As a swing roundabout gets faster, there is a force inwards that is keeping the swings from spinning off in a straight line. The swings are pulling out on the ropes with a centrifugal ("centre-fleeing") force.

Try whirling a bucket of water round your head fast (outdoors, just in case it spills). The water doesn't fall out as it is all pushed back by the centrifugal force.

Perhaps you can work out what forces are pushing on the parachute.

Making cream

People in industry use centrifugal force to separate different liquids such as milk and cream. Cream is less dense than milk, so it needs less centrifugal force to carry on going round in a circle. There is less centrifugal force in the centre, so the cream stays here and the milk is pushed out to the sides.

Terminal speed

Because of the forces of air resistance, something falling a long way, such as a parachutist, will accelerate to its terminal speed and will not be able to accelerate any more.

Machines, work and power

You use machines all the time to help you do things. Some are things you may not think of as machines, such as nutcrackers and tin-openers. Machines help you to do "work". Work has a special meaning in science – it is only done when something is moved. So although you may think you have "worked hard" for a test, you have in fact done very little work in the scientific sense. The amount of work done (in joules) is the force moved (in newtons) multiplied by the distance moved (metres).

Levers

Levers are a kind of simple machine. To a physicist, the ends of a see-saw are levers, and each person is trying to lift the other. A see-saw works best if the people are about the same weight and they sit at the ends. If one person is much heavier than the other, the heavier person needs to move nearer to the middle of the see-saw until the two people balance each other.

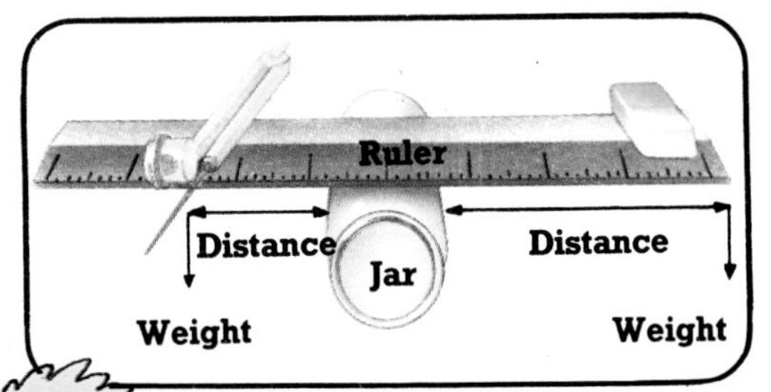

Balance two things, one heavier than the other, on a ruler that is balanced on the side of a jar. The heavier thing will have to be much nearer the middle, or pivot, of the see-saw. Calculate the work being done on each side of the see-saw. Does one side equal the other?

In order to have perfect balance, the work being done on each side of the see-saw must be equal. This means the distance from each person to the middle of the see-saw multiplied by that person's weight has to be the same.

All levers have three parts: the fulcrum, which is the pivot where movement takes place; the load arm, which is the length of lever between the load and the fulcrum; and the force arm, which is the length of lever between the applied force and the fulcrum.

A wheelbarrow is a simple lever system. The wheel is the fulcrum about which the load moves. The load placed in the wheelbarrow acts downwards, while the gardener lifts upwards on the handles. The load arm is measured from the wheel to the load.

If the force (effort) arm is four times as long as the load arm, the upward lift needed on the handles is equal to a quarter of the downward load or weight in the barrow. So the gardener can carry a much greater load in a wheelbarrow than he could carry in his arms.

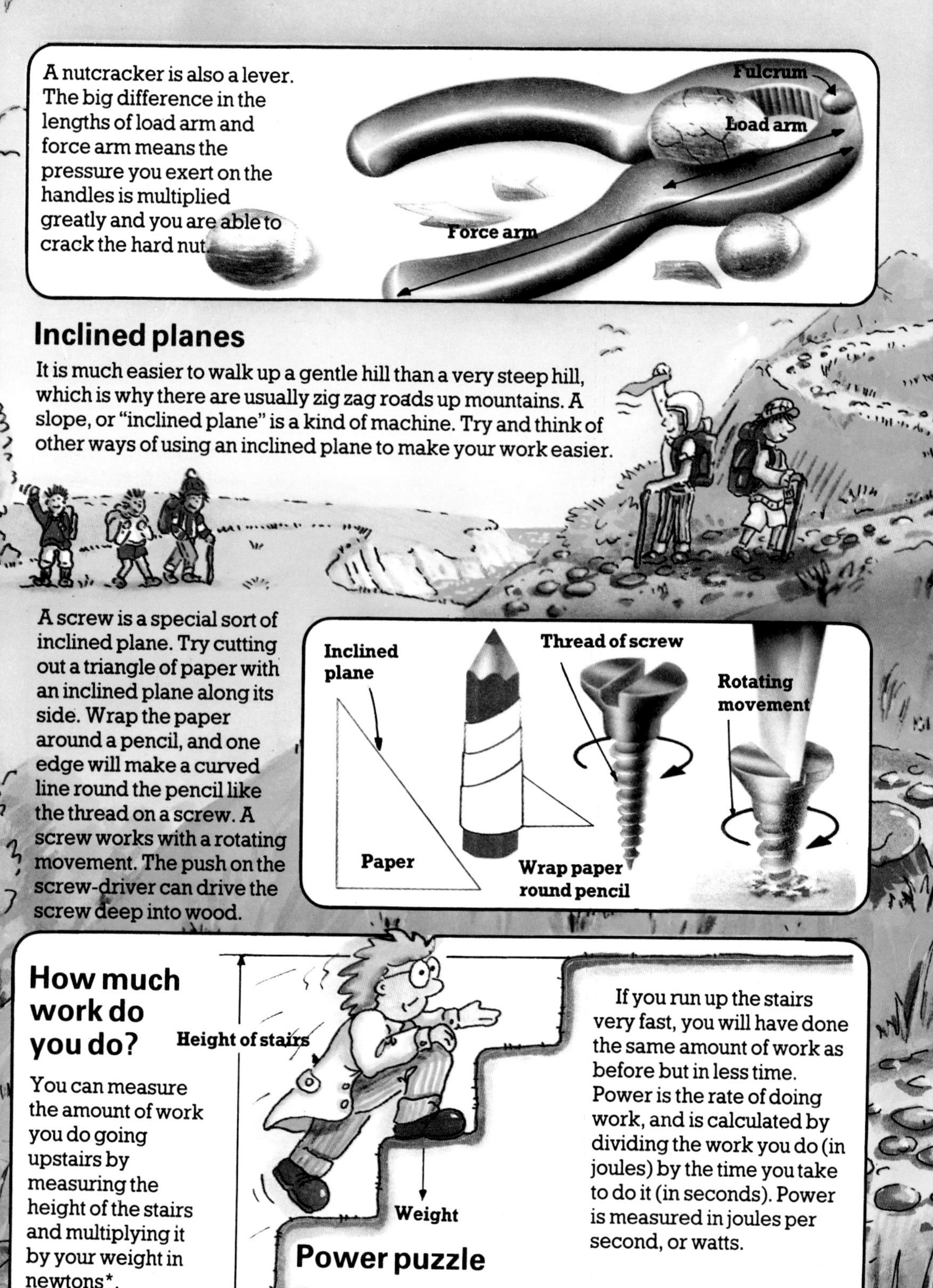

A nutcracker is also a lever. The big difference in the lengths of load arm and force arm means the pressure you exert on the handles is multiplied greatly and you are able to crack the hard nut.

Inclined planes

It is much easier to walk up a gentle hill than a very steep hill, which is why there are usually zig zag roads up mountains. A slope, or "inclined plane" is a kind of machine. Try and think of other ways of using an inclined plane to make your work easier.

A screw is a special sort of inclined plane. Try cutting out a triangle of paper with an inclined plane along its side. Wrap the paper around a pencil, and one edge will make a curved line round the pencil like the thread on a screw. A screw works with a rotating movement. The push on the screw-driver can drive the screw deep into wood.

How much work do you do?

You can measure the amount of work you do going upstairs by measuring the height of the stairs and multiplying it by your weight in newtons*.

If you run up the stairs very fast, you will have done the same amount of work as before but in less time. Power is the rate of doing work, and is calculated by dividing the work you do (in joules) by the time you take to do it (in seconds). Power is measured in joules per second, or watts.

Power puzzle

If you weigh 450 newtons and climb up stairs 10 metres high in 2 seconds, how much work and power have you used?

*See page 22 to find out about your weight in newtons.

Electricity and magnetism

Without electricity and magnetism, there would be no television, hi-fi systems, computers, video games, electric light and many of the other things around you. The next eight pages will tell you more about electricity and magnetism and how they work together.

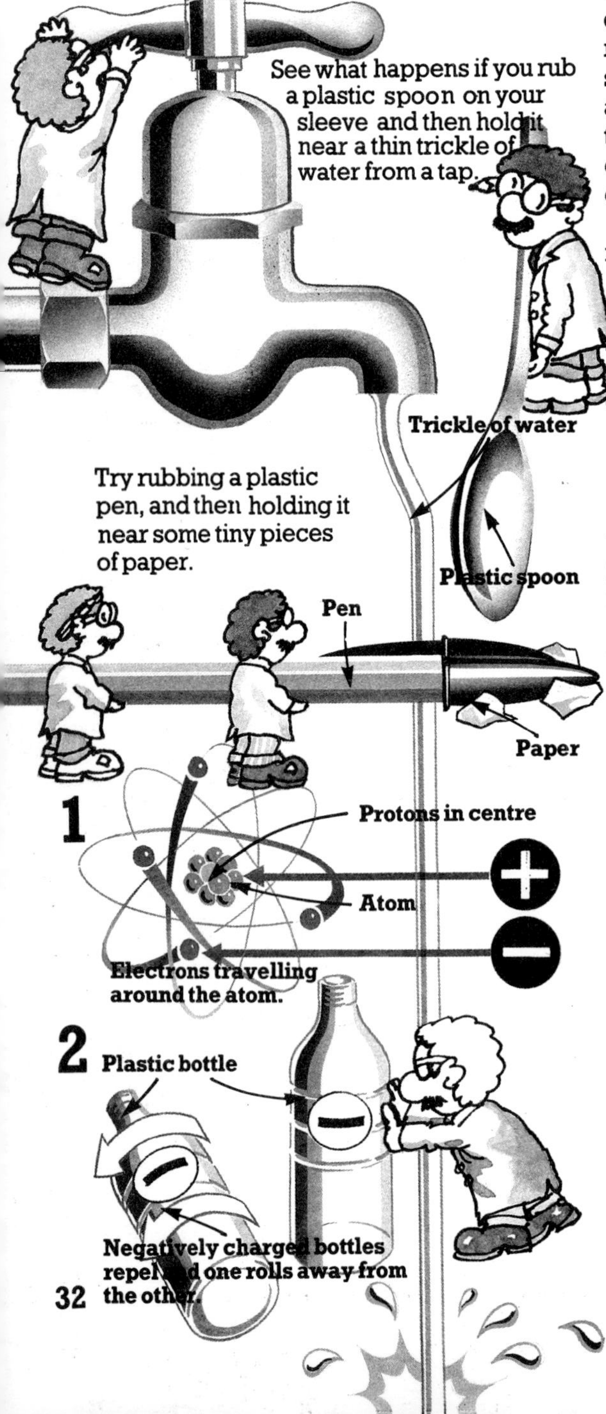

Static electricity

When you undress, you can sometimes hear a crackling noise as something nylon rubs against another material. If it is dark, you can sometimes see tiny flashes of electricity. This happens because of static electricity.

The Ancient Greeks knew that static electricity existed, but it was not until the eighteenth century that Benjamin Franklin realized that there are two unlike charges of static electricity, which he called positive and negative. He was the first to discover that storm clouds are charged with static electricity, and he invented the lightning conductor in 1752.

Because of these electric charges, almost magical things can happen.

If you sit on a chair, rub your rubber-soled shoes on the carpet and then touch something metal, you may feel a tiny electric shock. This is because electric charges are flowing through your body.

What is happening?

1. Everything is made up of atoms which themselves contain lots of charged particles. The positive charges are called protons and the negative charges electrons. In an uncharged (neutral) atom, the number of protons (+) is equal to the number of electrons (−). Electrons are much lighter than protons, and are on the edge of atoms, so they can move about. The protons are fixed in the centre or nucleus of the atom.

2. When two things such as wool and plastic are rubbed together, electrons sometimes move from one to the other. Rub two empty plastic bottles against wool. This makes them both negatively charged, with too many electrons. Put one on its side on a table, bring the second close to it, and the first bottle will roll away. Materials either negatively or positively charged can attract (pull) things that have an opposite charge. Materials having the same sort of charges repel (push) each other.

Charged things with uncharged things

What happens if you put something you have charged, like the pen below, next to something uncharged, like paper? If the pen has a negative charge it will repel the electrons on the paper nearest to it. This will leave the paper with a positive charge at one end and so it will be attracted to the pen and stick to it.

Eventually some of the extra electrons on the pen will travel through you to the ground to be "earthed". The paper will not be attracted to the pen any more and will fall off.

Negatively charged pen.

Paper eventually falls away from pen.

The protons on this side of the paper are attracted to the pen.

Negatively charged plastic bottle

Duck follows bottle

An electric question

Give an empty plastic bottle a negative charge by rubbing it with something woollen. Bring it close to a toy duck in the bath and the duck will follow behind the bottle. Why do you think this is? Find out on page 47. What would happen if you rubbed the duck with wool too?

**See page 34 for more about electric current and page 30 for more about work.*

Lightning

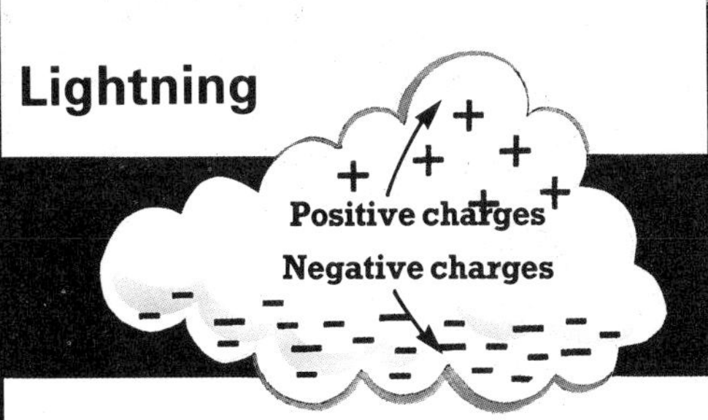

In thundery weather, clouds become charged by particles in them rubbing against each other. Positive and negative charges build up in different parts of the clouds until eventually a spark of negative charge leaps across from one side of the cloud to the other, making "sheet" lightning.

A very big charge on the cloud can "induce" an opposite charge on the Earth below. An electric current* then flows towards the Earth, making a flash of fork lightning. It lasts only a very short time, but a great amount of work* is done (about enough to run a 100W electric light bulb for a month). The air through which the current passes becomes very hot, but it returns to normal very quickly.

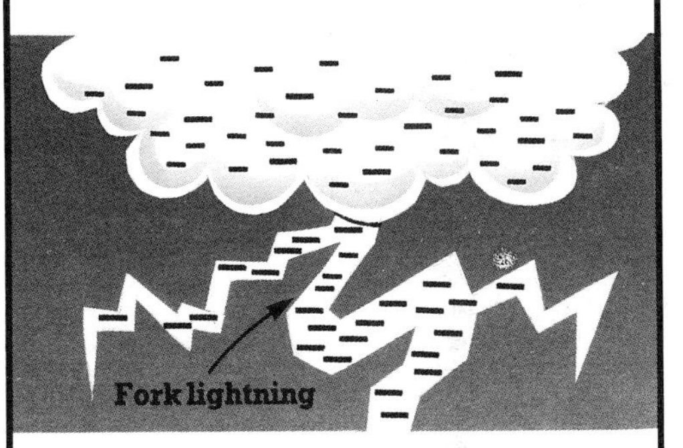

If the electric current hits anything on its way to Earth, it burns it. High buildings have lightning conductors on them – strips of highly conductive metal – which take the current safely down to Earth.

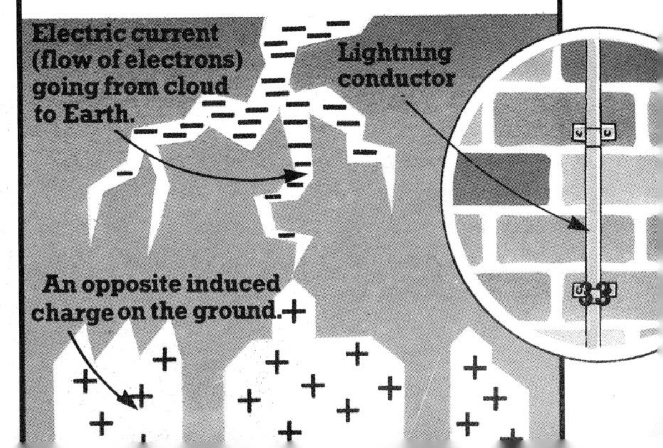

Current electricity

Static electricity means charges that stay still; they don't travel along wires or through the air. Current electricity is charges that are continuously on the move and it is this kind of electricity which makes things like light bulbs work. Power stations circulate electric current along the mains wires to the places that need current electricity.*

Conductors and insulators

Like heat, electricity travels better through some materials than others. Good conductors of electricity have many more "free" electrons than insulators. Under normal conditions these electrons drift to and fro from atom to atom in a random way. Metals have lots of free electrons, which is why they are good conductors.

Copper (good conductor) inside.

Insulator

If you look at a piece of electrical flex you will find two copper wires (good conductors) wrapped in rubber (an insulator) in order to insulate the wires and make them safe.

WARNING!

The electricity in your home is very dangerous. Never touch the metal prongs in plugs, because electric current could flow through the prongs and through you. It could give your heart such a shock that it could stop beating.

Whenever there are more electrons at one end than at the other end of a conductor, free electrons in the conductor are forced to move one way, as an electric current, towards the end with fewer electrons. The difference in electrons between one end and the other is called the potential difference and is measured in volts. A battery can set up a potential difference. Current is a measure of the number of electrons drifting along the wire, and is measured in amperes (or amps).

How a battery works

Inside a battery is a special chemical paste called the electrolyte that can conduct electricity. It is made of billions of positive and negative particles. The case of the battery is made of zinc and a carbon rod lies in the electrolyte. The zinc and carbon are both electrodes. A chemical reaction in the electrolyte sends positive particles to one electrode (the less reactive one) and negative particles to the other.

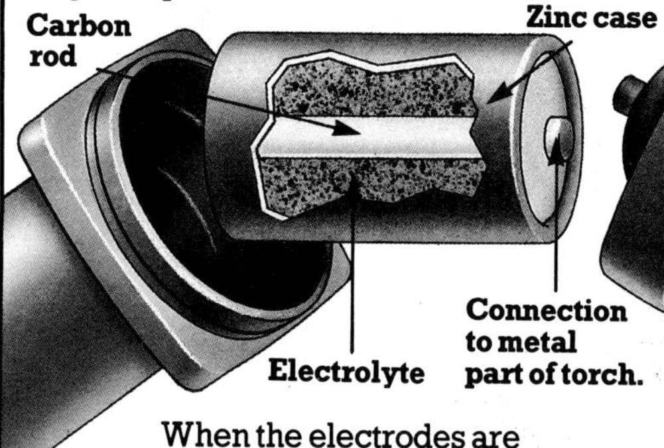

When the electrodes are connected by touching metal parts of a torch, electric current flows. When the electrolyte is used up, the current cannot flow any more and the battery is "dead".

Make your own battery

Drawing pin (brass)

Paper clip (steel)

Copper wire

Stick two pieces of metal that are different into a lemon, making sure they are not touching each other. Wrap some copper wire around the ends of the metals, and connect the other ends of the wire to a 1.5 volt torch bulb which is in a lampholder. The lamp may light up. This is because the metals are acting as electrodes, and the lemon as the electrolyte.

**On page 42 there is a computer program which you can run to work out how much electricity you use at home, and how much your electricity bill will be.*

Electrical resistance

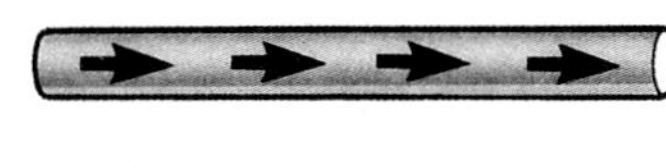

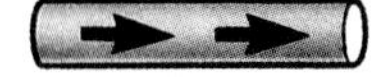

Good conductors of electricity allow electrons to flow easily. Sometimes, though, they bump into atoms in the wire and this slows them down. This braking effect is called the wire's resistance. The longer the piece of wire, the more resistance it has.

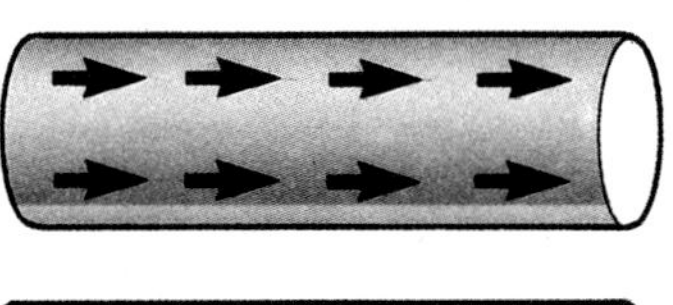

A thick wire has a lower resistance than a thin wire. There is a larger area of wire for the electrons to pass along. It is a bit like a motorway that can carry more traffic than a single-lane country road.

Electric light

The wire in an electric light bulb is made of very thin, coiled tungsten metal. (Tungsten is used because it won't melt unless it gets very hot.) The electrons hit atoms in the wire, making them vibrate more and more and get hotter and hotter, until the wire glows with the "white heat" which we see as light.

Electrons passing along the very fine wire keep bumping into atoms. The atoms vibrate, and give off heat and light.

Glass bulb

Wires support the filament

Coiled filament

A number in watts (W) telling you how much power the bulb uses and how brightly it glows. The bigger the number the brighter the glow, and the more electricity it uses.

The bulb is filled with a non-reactive gas, such as argon. If the bulb contained ordinary air, the metal would combine with oxygen and burn up.

Because the wire is tightly coiled, more wire can be put in the bulb, and more light is produced.

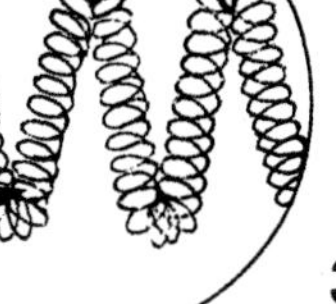

Close up of coiled tungsten wire (called the filament).

Two sorts of electric current

The electric current a battery makes is called direct current (d.c.). It only flows in one direction. But the electric current that comes from power stations travels as "alternating current" (a.c.) which changes direction a hundred times a second. Alternating current can be transformed (using a "transformer") into higher voltages before it is sent over long distances. At high voltages less energy is lost as heat on the way.

Magnetism

Magnets are very useful to us. They are an essential part of loudspeakers, microphones, electric motors, door bells and many other things.

Magnetism was first discovered 2,500 years ago in a stone called lodestone, which people used to make the first compasses. Only the metals iron, nickel and cobalt can be magnetized on their own, but powerful magnets can be made by mixing these with other metals. Steel is iron with a little carbon in it, so it makes strong magnets too. See if you can find a magnet and find out what sorts of things it can pick up.

Destroying magnetism

When something is magnetized, many of its molecules are pointing the same way. To destroy magnetism, you need to mix up all the "molecular magnets" again. You can do this by hitting a magnet with a hammer, or heating it until it is red-hot and let it cool down (don't do this yourself).

What is a magnet?

Imagine lots of matches representing groups of molecules* in a magnetic substance. Each matchstick is like a little magnet with a north pole at its head end and a south pole at the other end.

A bar of unmagnetized iron can be thought of as a jumble of matchstick magnets, so mixed up that all their magnetic forces cancel out.

When the iron bar is magnetized, many of the molecular magnets line up, with their north poles pointing the same way.

Fields of force

Iron filings

You can't see what makes magnets work, but there are magnetic forces around them and you can see the pattern they make by using iron filings. Put about one teaspoonful of filings into a box and shake it around so they cover the bottom. Hold the box just above a magnet, tap the box and you will see the filings jump up. They fall down into a pattern made up of curved lines. This pattern is called a field of force. These lines of force show what happens in the space round a magnet. Try doing this with two magnets, their like poles together.

Iron filings

Make a magnet

Playing with two bar magnets, you will find that a north pole will attract (pull) a south pole. Two "like poles" – south/south or north/north – repel (push) each other. You can magnetize a steel needle by stroking it in the same direction eight or nine times with the north pole of a bar magnet. As you do this, the north pole of the magnet pulls the south poles of the tiny molecular magnets in the needle and makes them start to line up. See if the needle can pick things up now. If not, stroke it some more.

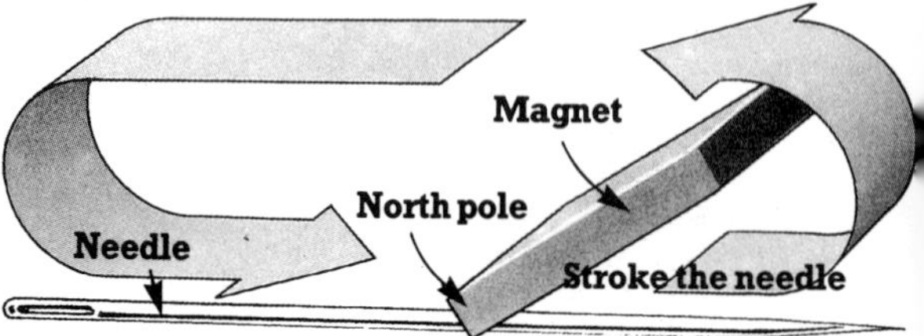

**See page 14 for more about molecules.*

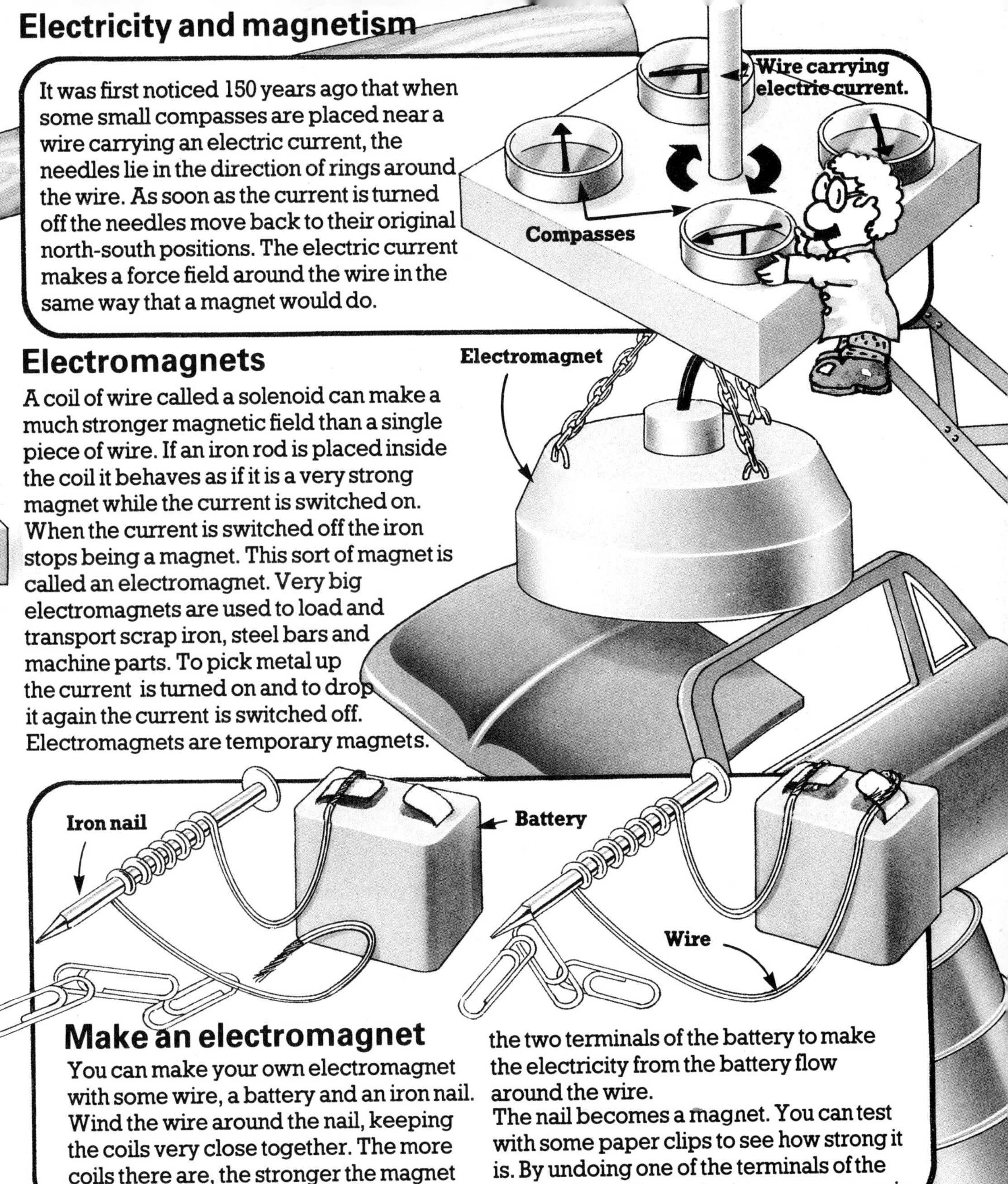

Electricity and magnetism

It was first noticed 150 years ago that when some small compasses are placed near a wire carrying an electric current, the needles lie in the direction of rings around the wire. As soon as the current is turned off the needles move back to their original north-south positions. The electric current makes a force field around the wire in the same way that a magnet would do.

Electromagnets

A coil of wire called a solenoid can make a much stronger magnetic field than a single piece of wire. If an iron rod is placed inside the coil it behaves as if it is a very strong magnet while the current is switched on. When the current is switched off the iron stops being a magnet. This sort of magnet is called an electromagnet. Very big electromagnets are used to load and transport scrap iron, steel bars and machine parts. To pick metal up the current is turned on and to drop it again the current is switched off. Electromagnets are temporary magnets.

Make an electromagnet

You can make your own electromagnet with some wire, a battery and an iron nail. Wind the wire around the nail, keeping the coils very close together. The more coils there are, the stronger the magnet will be. Twist the ends of the wire around the two terminals of the battery to make the electricity from the battery flow around the wire.
The nail becomes a magnet. You can test with some paper clips to see how strong it is. By undoing one of the terminals of the battery the nail stops being a magnet again.

Another way of making magnets

A magnet can sometimes make something else into a magnet without even touching it. Its lines of force stretch out across space and make the "magnet-molecules" in the object line up. This is called magnetic induction.

Electric motors

Imagine a wire carrying an electric current, placed between two magnets. The magnetic fields from the magnet interact with the electrical magnetic field from the wire. The force that results moves the wire to a new position. Electric motors use this idea.

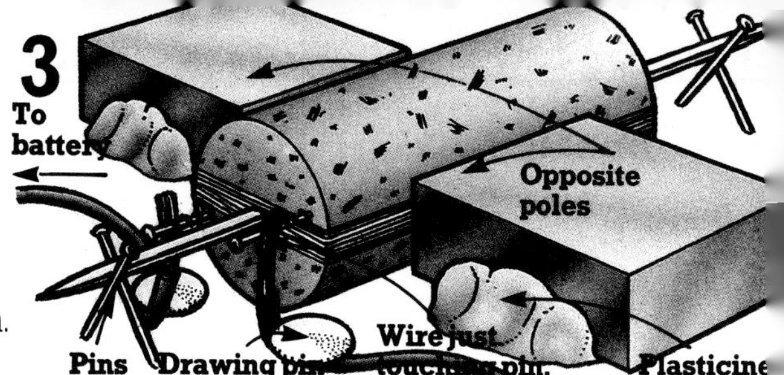

Make an electric motor

It may help you to understand electric motors better if you build one yourself. You will need:
2 permanent magnets
1 large cork
6 pins
1 knitting needle
some thin, insulated copper wire
plasticine
soft board (fibreboard)
4½ volt battery
2 pieces of thicker, insulated copper wire
a sharp knife
2 drawing pins

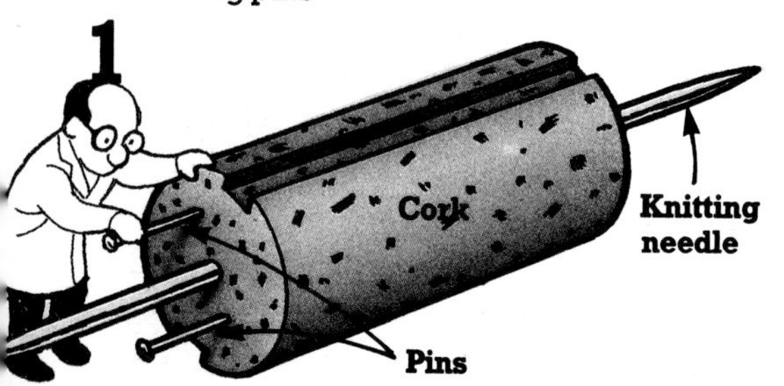

Cut a narrow channel on either side of the cork. Push the knitting needle through the centre of the cork and push two pins into one end of the cork.

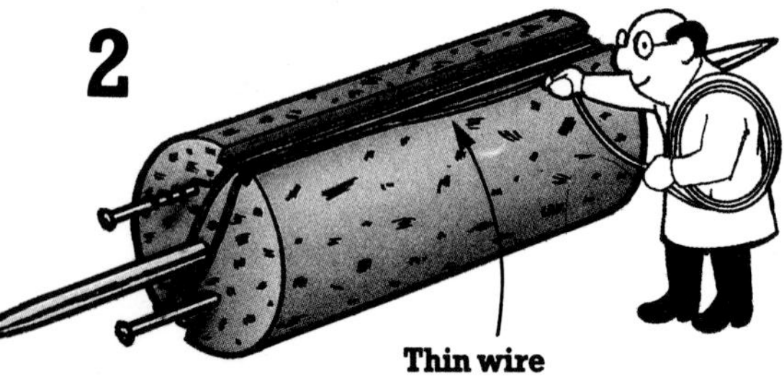

Strip about 2cm of the insulation off one end of the thin wire. Wrap it round one of the pins. Wind the wire round the cork about 30 times. Strip the insulation off the other end and wind it round the other pin.

Push two pairs of pins into the board so the knitting needle can rest in them like a cradle. Strip the insulation from the ends of the thicker copper wires. Use drawing pins to hold them so they just touch the pins in the cork.

Use plasticine to support the magnets on each side of the coil, with opposite poles facing. Connect the wires to a 4½ volt battery and give the cork a flick to start it going round.

What is happening?

There are two separate fields of force working together in the motor. These diagrams show what happens to the field pattern. Imagine the wire carrying the current is sticking straight out of the page towards you.

The two magnets, with their opposite poles together, set up fields of force that cross in the space between them, like this:

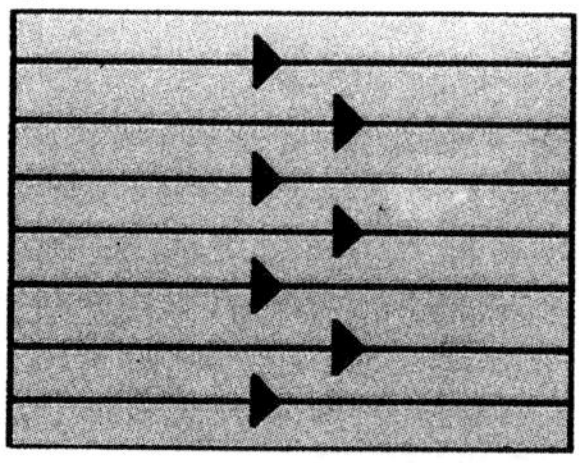

The wire makes its own field of force, like this:

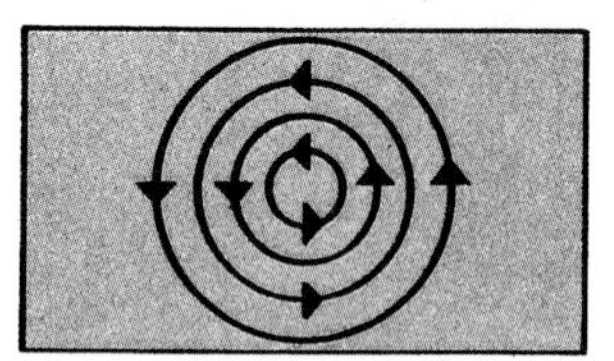

The combined force looks like this. It has a "catapult" effect on the wire, pushing it to one side. In the motor this has the effect of pushing one side of the coil up and the other side down, which means the coil will rotate.

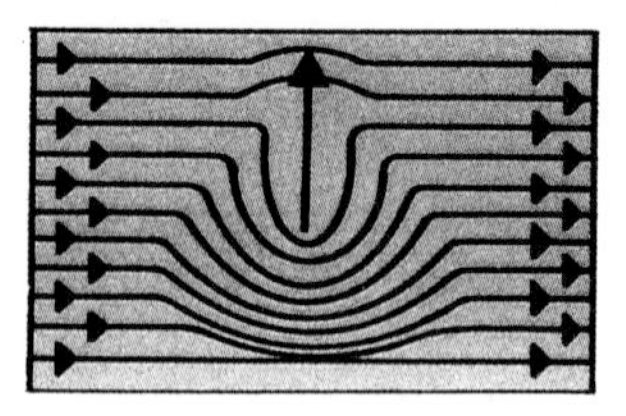

Electric motors are used to do many useful jobs of work; in vacuum cleaners, drills, trains, lifts and washing machines for instance. The motor is using electrical energy to do "work" (i.e. drive a machine).

How loudspeakers work

Loudspeakers use a combination of magnetic and electric fields to bring speech and music into your home, and to carry your voice over the telephone. They convert electrical energy into sound energy.

A loudspeaker contains a movable coil of wire, attached to a large cone. The coil fits loosely over the centre of a cylindrical permanent magnet so that the coil is in a strong magnetic field.

Varying electric currents pass through the coil of wire. Because of the catapault effect (as in the electric motor), the coil moves. The cone is connected to the coil so it moves too, sending out vibrations (sound waves) into the air. The vibrations vary with the current.

It is the combination of the magnetic and electric fields that produces the movement of the coil and the cone, to send out the sound waves.

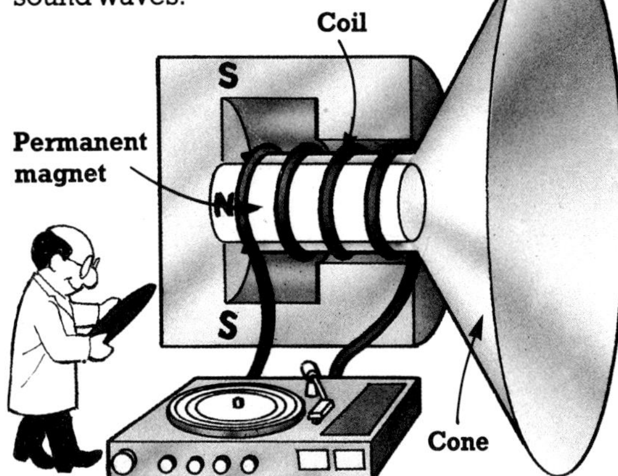

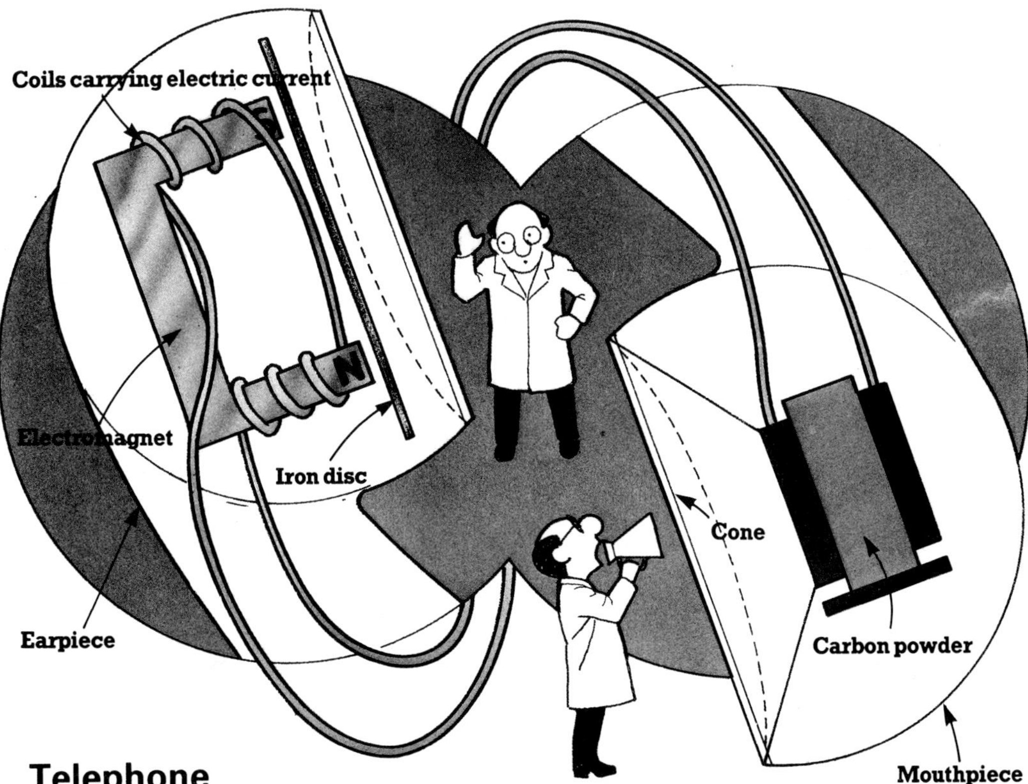

Telephone

Here the varying electric currents pass round the coils of an electromagnet, which attracts an iron disc. As the currents vary, the movement of the disc varies and makes a sound wave in the air.

The varying electric currents are produced by a carbon microphone in the mouthpiece of a telephone. The sound waves force a cone in and out. This squeezes some carbon powder, through which the electric current is flowing. When the carbon is squeezed, its resistance is less and so the current passing through it changes as the sound wave changes.

Electromagnetic spectrum

You have already found that light energy travels in electromagnetic waves. But there is a wide range of other electromagnetic waves too. Together they make up the electromagnetic spectrum. All these waves travel at the same speed – the speed of light, which is 300 million metres per second. The difference between them is in their varying wavelengths and in the way they affect things.

Gamma rays

Gamma rays have the tiniest wavelength of all the electromagnetic waves. They are given off by some substances, for example uranium, which are radioactive. Radioactive substances are constantly giving out energy from the nuclei of their atoms, either in the form of particles, or as gamma rays. These rays are very penetrating, they can even travel through cement and lead. They can be very dangerous because they damage the cells of our bodies.

X-rays

These were discovered accidentally in 1895 by a German physicist called Röntgen. He called them "X-the- unknown" because he didn't fully understand them. To produce X-rays, a beam of electrons is fired at a heavy target, usually made of tungsten.

Your body tissue is mostly made up of hydrogen, oxygen, carbon and nitrogen, but your bones contain calcium, which is denser and absorbs rays better. When X-rays are shone through your body most of them go right through and fall onto a photographic plate on the other side, but where there are bones the rays are stopped and this makes a shadow on the plate. From this doctors can tell if a bone is broken or out of place. They can also see accidentally swallowed objects.

Ultraviolet waves

These are beyond the violet end of the visible light spectrum; we can't see them, but most insects can. They usually come from the Sun and most of them are absorbed by the layer of ozone that surrounds the Earth. Ultraviolet waves make you sun-tanned, but if you stay in the Sun too long they are dangerous, because you get sunburnt. Scientists have now invented "sun-beds" which can produce ultraviolet light artificially.

Visible light

Turn to page 6 to find out more about visible light.

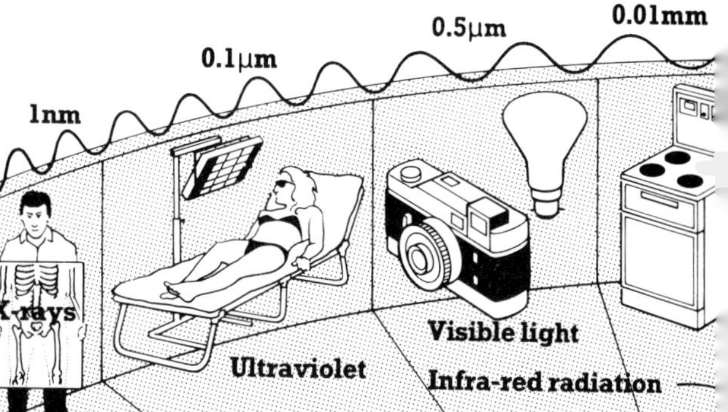

Infra-red radiation

Infra-red radiation has longer wavelengths than red light. We cannot see it, but we feel it as heat. We call it heat radiation because it is given off by most hot things. Only the infra-red rays near the visible spectrum pass through glass, the longer ones are absorbed by it.

Microwaves

Microwaves have wavelengths between 1mm and 0.3m, which is between infra-red radiation and radio waves. Radar, which is a way of locating a distant object, uses microwaves. The waves are fired at the object, and some are reflected back. From the time it takes for them to come back, it can be worked out how far away the object is and how far it is moving.

Microwave "ovens" are used to heat and cook food very quickly. The waves give the molecules in the food lots of energy, making the food so hot that even a potato can be baked in 4 minutes.

TV and radio waves

Radio waves can be used to carry messages and television pictures around the world at the speed of light. Radio waves are grouped into bands and each band has a special set of uses. Cameras and microphones create electronic signals which are combined with radio waves and sent out to be picked up by aerials connected to people's TV sets.

Nowadays more and more TV programs are being transported through underground cables, rather than across space on electromagnetic waves. It is possible to send many more channels by cable without them interfering with each other.

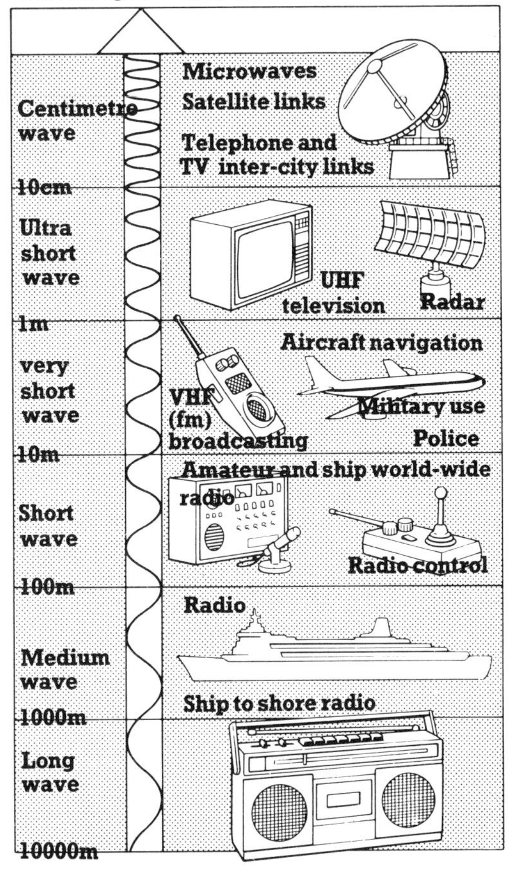

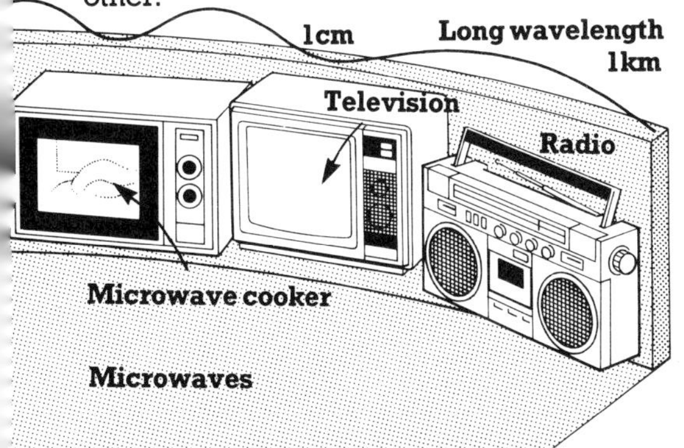

Lasers

A laser is any device that emits beams of laser light. This is an unusual sort of light because, unlike other beams of electromagnetic waves, it covers a very small range of wavelengths, and it is polarized. Usually electromagnetic waves vibrate in all directions at right angles to the direction of the wave. When light is polarized the vibrations are in one direction only. The laser emits polarized light, and it is called a coherent source of light. (Why do you think some sunglasses are called polarized?).

The most common source of laser light is a crystal, such as a ruby, which can be stimulated by a flash of very bright light. A special mixture of gases can also give off laser light when an electric current passes through it.

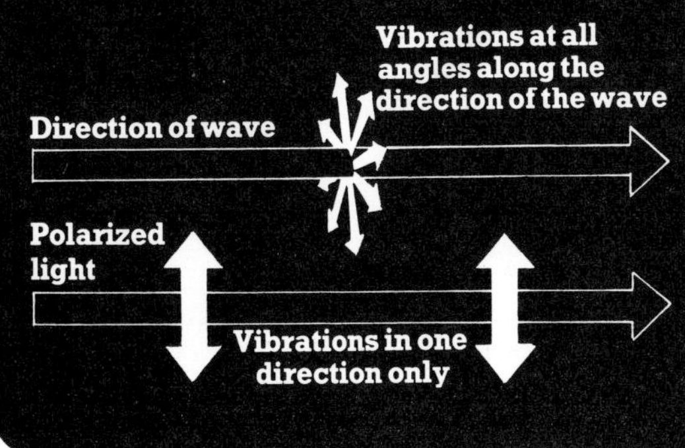

Fibre optic cables

The latest development in cable systems is the use of fibre optic cables. These are made of bundles of very thin glass, down which is passed a pattern of laser light. Sound can be converted into patterns of laser light and transmitted in this way across very long distances.

Home electricity computer program

This program works out how much electricity things like your television and cooker use. You can also calculate how much your three-monthly bill should add up to.

If you have, or can borrow, a BBC microcomputer, you can run the program as it is. Lines that need changing for other computers are marked with symbols and printed at the end of the program. Each symbol corresponds to a different computer. They are:

▲ VIC and PET
▮ ZX SPECTRUM, ZX81
● APPLE
■ TRS-80
○ ORIC

Before you can work out your bill you need to check on a recent electricity bill or ring up, to find out the unit price (UP) you pay for electricity.

Different computers vary too much to be able to give general instructions for graphics, but there is an example of a graphics subroutine for a light bulb, which will work on a Spectrum (Timex 2000). This should be added at the end of the program, and another line put into the program to call it up: 1675 GOSUB 3000.

You could perhaps try writing your own graphics routines for the other appliances.

```
 10 REM INITIALISE
 20 LET N=10: REM NO. OF APPLIANCES
 30 DIM U(N): REM UNITS USED
▮40 DIM A$(N): REM NAMES
 50 LET TU=0: REM POWER USED
 60 LET UP=2.5: REM UNIT PRICE
 70 LET A$(1)="COOKER"
 80 LET A$(2)="IMMERSION HEATER"
 90 LET A$(3)="FAN HEATER"
100 LET A$(4)="RADIANT HEATER"
110 LET A$(5)="LIGHT BULB"
120 LET A$(6)="WASHING MACHINE"
130 LET A$(7)="TELEVISION"
140 LET A$(8)="RADIO"
150 LET A$(9)="CONVECTOR HEATER"
160 LET A$(10)="HI-FI STEREO"
170 REM * PRINT INTRO PAGE *
180 CLS
190 PRINT
200 PRINT
210 PRINT "ELECTRICITY BILL"
220 PRINT "  CALCULATION"
230 PRINT "  ==========="
240 PRINT
250 PRINT "POWER"
260 PRINT "STATION >>>>>>>>>>"
270 PRINT "                 TRANS-"
280 PRINT "                 FORMER"
290 FOR I=1 TO 4
300 PRINT "                    V"
310 NEXT I
320 PRINT "                  HOUSE"
330 PRINT
340 PRINT"PRESS SPACE TO START"
350 GOSUB 810
360 REM MAIN MENU PAGE
370 CLS
380 PRINT "CHOOSE THE APPLIANCE"
390 PRINT "THAT YOU WANT TO ENTER"
400 PRINT "NEXT, OR TYPE 0 TO"
410 PRINT "CALCULATE YOUR BILL"
420 PRINT
```

```
430 PRINT "                    UNITS"
440 FOR I=1 TO N
450 IF U(I)>0 THEN PRINT ;I;" ";A$(I);
    TAB (19);U(I)
460 IF U(I)=0 THEN PRINT ;I;" ";A$(I)
470 NEXT I
480 PRINT
490 PRINT "TYPE A NUMBER AND THEN"
500 PRINT "PRESS ENTER";
510 INPUT C
520 IF C<0 OR C>N THEN GOTO 360
530 IF C=0 THEN GOTO 580
540 CLS
550 PRINT
560 ON C GOSUB 1060,1280,1330,1530,1650,
    1700,1900,2060,2110,2160
570 GOTO 360
580 REM FINAL PAGE
590 CLS
600 FOR W=1 TO N
610 LET TU=TU+U(W)
620 NEXT W
630 PRINT
640 PRINT "ELECTRICITY BILL"
650 PRINT "    ESTIMATE"
660 PRINT "    ========"
670 PRINT "(FOR 3 MONTHS)"
680 PRINT
690 PRINT "UNITS USED :"
700 PRINT ;TU;" KILOWATT-HRS"
710 PRINT
720 PRINT "UNIT PRICE :";UP;" PENCE"
730 LET TC=(INT(UP*TU))/100
740 PRINT
750 PRINT
760 PRINT "TOTAL DUE : ‘";TC
770 PRINT
780 PRINT "PRESS SPACE TO RUN AGAIN"
790 GOSUB 810
800 RUN
810 LET I$=INKEY$(0)
820 IF I$<>" " THEN GOTO 810
830 RETURN
840 REM PAGE FOR INPUT
850 CLS
860 PRINT
870 PRINT N$
880 PRINT
890 PRINT ;P*1000;" WATTS"
```

```
900 FOR I=1 TO 7
910 PRINT
920 NEXT I
930 PRINT "HOW LONG IS THIS APPLIANCE"
940 PRINT "USED EACH WEEK, ON AVERAGE?"
950 PRINT "(IN HOURS)"
960 PRINT "TYPE THE NUMBER THEN"
970 PRINT "PRESS RETURN";
980 INPUT T
990 LET U(C)=U(C)+P*T*13
1000 RETURN
1010 REM MOVE DOWN 5 LINES
1020 FOR X=1 TO 5
1030 PRINT
1040 NEXT X
1050 RETURN
1060 REM * COOKER *
1070 PRINT A$(C)
1080 GOSUB 1010
1090 PRINT "PRESS  1) FOR RING"
1100 PRINT "       2) FOR OVEN"
1110 PRINT "       3) FOR GRILL"
1120 PRINT
1130 INPUT I
1140 IF I<1 OR I>3 THEN GOTO 1130
1150 ON I GOTO 1160,1200,1240
1160 LET N$="COOKER RING"
1170 LET P=1
1180 GOSUB 840
1190 RETURN
1200 LET N$="COOKER OVEN"
1210 LET P=3
1220 GOSUB 840
1230 RETURN
1240 LET N$="COOKER GRILL"
1250 LET P=1.5
1260 GOSUB 840
1270 RETURN
1280 REM * IMMERSION HEATER *
1290 LET N$=A$(C)
1300 LET P=3.5
1310 GOSUB 840
1320 RETURN
1330 REM * FAN HEATER *
1340 LET N$="FAN HEATER"
1350 PRINT N$
1360 GOSUB 1010
1370 PRINT "IS IT  1) FULL ON"
1380 PRINT "       2) HALF ON"
```

```
1390 PRINT "        3) COLD AIR"
1400 INPUT I
1410 IF I<1 OR I>3 THEN GOTO 1400
1420 ON I GOTO 1430,1460,1490
1430 LET N$=N$+" (FULL ON)"
1440 LET P=3
1450 GOTO 1510
1460 LET N$=N$+" (HALF ON)"
1470 LET P=1.5
1480 GOTO 1510
1490 LET N$=N$+" (COLD AIR)"
1500 LET P=0.3
1510 GOSUB 840
1520 RETURN
1530 REM * RADIANT HEATER *
1540 LET N$="RADIANT HEATER"
1550 PRINT N$
1560 GOSUB 1010
1570 PRINT "ARE YOU USING "
1580 PRINT "1,2 OR 3 BARS"
1590 INPUT I
1600 IF I<1 OR I>3 THEN GOTO 1590
1610 LET N$=N$+" ("+STR$(I)+" BARS)"
1620 LET P=I
1630 GOSUB 840
1640 RETURN
1650 REM * LIGHT BULB *
1660 LET N$=A$(C)
1670 LET P=0.1
1680 GOSUB 840
1690 RETURN
1700 REM * WASHING MACHINE *
1710 LET N$="WASHING MACHINE"
1720 PRINT N$
1730 GOSUB 1010
1740 PRINT "IS IT  1) WASHING"
1750 PRINT "        2) SPINNING"
1760 PRINT "        3) HEATING"
1770 INPUT I
1780 IF I<1 OR I>3 THEN GOTO 1770
1790 ON I GOTO 1800,1830,1860
1800 LET N$=N$+" (WASHING)"
1810 LET P=0.8
1820 GOTO 1880
1830 LET N$=N$+" (SPINNING)"
1840 LET P=0.8
1850 GOTO 1880
1860 LET N$=N$+" (HEATING)"
1870 LET P=3
1880 GOSUB 840
1890 RETURN
1900 REM * TELEVISION *
1910 LET N$="TELEVISION"
1920 PRINT N$
1930 GOSUB 1010
1940 PRINT "IS IT  1) COLOUR"
1950 PRINT "    OR 2) BLACK AND WHITE"
1960 INPUT I
1970 IF I<1 OR I>2 THEN GOTO 1960
1980 IF I=2 THEN GOTO 2020
1990 LET N$=N$+" (COLOUR)"
2000 LET P=0.4
2010 GOTO 2040
2020 LET N$=N$+" (BLACK AND WHITE)"
2030 LET P=0.3
2040 GOSUB 840
2050 RETURN
2060 REM * RADIO *
2070 LET N$=A$(C)
2080 LET P=0.05
2090 GOSUB 840
2100 RETURN
2110 REM * CONVECTOR HEATER *
2120 LET N$=A$(C)
2130 LET P=3
2140 GOSUB 840
2150 RETURN
2160 REM * HI-FI STEREO *
2170 LET N$=A$(C)
2180 LET P=0.15
2190 GOSUB 840
2200 RETURN
```

Light bulb graphics

Below is a graphics subroutine for a light bulb. It will only work on a Spectrum (Timex 2000) and should be added here. In order to call it up another line must be put into the program: 1675 GOSUB 3000.

```
3000 REM GRAPHICS FOR LIGHT BULB
3010 CLS : PLOT 175,40: DRAW 0,32:
     DRAW -8,32,.7: DRAW 48,0,-4.9:
     DRAW -8,-32,.7: DRAW 0,-32
3020 PLOT 184,40: DRAW -8,88,.2
3030 PLOT 199,40: DRAW 8,88,-.2
3040 PRINT AT 5,22; INK 6; BRIGHT 1;"****"
3050 RETURN
```

Using different computers

Below is a list of changes that will enable you to run this program on other computers too. The symbols on the left-hand side of the column correspond to different computers. These instructions need to be inserted into the program in the relevant places.

```
▮    40 DIM A$(10,16)
▮   560 GOSUB 1060*(C=1)+1280*(C=2)+1330*(C=3
    )+1530*(C=4)+1650*(C=5)+1700*(C=6)+1900*(C=
    7)+2060*(C=8)+2110*(C=9)+2160*(C=10)
○   810 LET I$=KEY$
▲   810 GET I$
●   810 LET I$=""
●   812 IF PEEK(-16384)>127 THEN GET I$
▮■  810 LET I$=INKEY$
▮  1150 GOTO 1160*(I=1)+1200*(I=2)+1240*(I=3)
▮  1420 GOTO 1430*(I=1)+1460*(I=2)+1490*(I=3)
▮  1790 GOTO 1800*(I=1)+1830*(I=2)+1860*(I=3)
```

Physics words

This is a selection of some of the most important physics words and laws. Some you will have met in this book already. You will find that they are useful to all sorts of people, not just students: computer engineers, mechanics, electricians, space scientists, radiographers, sound engineers and many other people need to know a bit of physics for their jobs.

Acceleration. The increase of velocity every second measured in metres per second per second, (m/s^2).

Alternating current (a.c.). An electric current that flows first in one direction, then in the other.

Ampere (amp). The quantity of electric current flowing every second.

Amplitude. The height of a wave.

Archimedes Principle. A body floating in a fluid displaces a weight of fluid equal to its own weight.

Atom. The smallest particle of an element that can take part in a chemical reaction.

Centre of gravity. The point at which the weight of a body appears to act.

Conductor. A substance or body that offers a relatively small resistance to the passage of an electric current, or to heat.

Conservation of energy (law of). In any closed system energy cannot be created or destroyed, although its form may be changed.

Convection. The transfer of heat in a fluid (or gas) by the movement of the fluid itself.

Coulomb. The unit of electric charge. One coulomb is the charge transported in one second by an electric current of one ampere.

Critical angle. The angle of incidence of light proceeding from a denser medium to a lighter one at which grazing incidence occurs (angle of refraction = 90°).

Current. The rate of flow of electricity, measured in amperes.

Decibel (dB). The unit of the intensity of sound.

Density. The mass per unit volume of a substance, often measured in kilograms per cubic metre (kg/m^3).

Direct current (d.c.). An electric current that flows in one direction only.

Electron. A negatively charged particle, present in all atoms. Free electrons are responsible for electrical conduction in most substances.

Energy. A measure of the capacity to do work, measured in joules (J).

Force. Any action that alters a body's state of rest or of uniform motion in a straight line. Measured in newtons (N).

Frequency. The number of waves or cycles in one second.

Friction. A force that occurs whenever two things rub together.

Gravity. The pulling force of the Earth.

Incidence (angle of). The angle between the ray striking a surface and the normal to the surface at the point of incidence.

Inertia. The property of a body where it persists in a state of rest or uniform motion in a straight line.

Insulator. A substance that provides high resistance to the passage of an electric current, or to heat.

Joule (J). The unit of energy. One joule is the work done when a force of 1 newton moves something through 1 metre distance.

Kinetic energy (K.E.). The energy of movement, measured in joules.

Mass. The amount of matter in a body measured in kilograms.

Momentum. The mass of a body multiplied by its velocity.

Motor. A machine that converts electrical energy into mechanical energy.

Newton (N). The unit of force. One newton provides a mass of 1 kg with an acceleration of 1 m/s^2.

Newton's laws of motion
1. Every body remains in a state of rest or uniform motion in a straight line unless acted upon by forces from the outside.
2. The amount of acceleration of a body is proportional to the acting force and inversely proportional to the mass of the body.
3. Every action has an equal and opposite reaction.

Ohm (Ω). The unit of resistance.

Pascal (Pa). The unit of pressure. One pascal is the pressure that results from a force of one newton acting on an area of one square metre. One pascal is equal to one newton per square metre (N/m^2).

Pitch. The pitch of sound depends on its frequency. High frequency = high pitch, low frequency = low pitch.

Potential difference (p.d.). The voltage in part of an electric circuit, measured in volts.

Potential energy (P.E.). Energy which is stored up, ready to be used, measured in joules (J).

Power. The rate energy is expended, or work is done. Power is measured in watts.

Pressure. The force per unit area. Measured in pascals, or newtons per square metre (N/m^2) or millimetres of mercury (mmHg).

Proton. A positively charged particle that is present in all nuclei.

Radiation. Any form of energy that moves as waves, rays or a stream of particles.

Reflection (laws of)
1. The incident ray, normal and the reflected ray all lie in the same plane.
2. The angle of incidence is equal to the angle of reflection.

Refraction. The change of direction that a ray undergoes when it enters another transparent substance.

Resistance. The more resistance a wire has, the less current it can carry. Resistance is measured in ohms.

Transformer. A device that alters the alternating voltage of an electric current.

Velocity. The speed of a body in a particular direction measured in metres per second (m/s).

Volt. One volt is the force needed to carry one ampere of current against one ohm of resistance.

Watt. The unit of power. One watt is the power resulting from the dissipation of one joule of energy in one second. In electricity watts = amps × volts.

Wavelength. Length from one peak in a series of waves to the next.

Weight. The force exerted on matter by the gravitational pull of the Earth, measured in newtons (N).

Important physics equations
Force (N) = mass (kg) × acceleration (m/s^2)
Volts = current (amps) × resistance (ohms)
Velocity of a wave (m/s) = frequency × wavelength (m)
Pressure (N/m^2) = force (N) × area (m^2)
Power (watts) = volts × current (amps)

Answers

Page 5: Energy puzzle

1. Here the dog has potential chemical and gravitational energy.
2. When running downstairs, the dog is changing potential energy to kinetic (movement) energy.
3. At the bottom of the stairs, the food the dog eats is replacing some of the chemical potential energy that changed to kinetic energy when he came downstairs.

Page 6: Light energy

The Sun, a torch and a candle are sources of light, the other things are only reflecting light that comes from another source. Even the Moon is only reflecting the Sun's light.

Page 20

The flute produces music by blowing.
The piano has little hammers inside that hit the strings.
The violin and harp both have strings that are plucked.

Page 22

If your mass is 60kg you would weigh 100N on the Moon, and your mass would be the same on the Moon as on the Earth.

Page 24

The water molecules on the surface of the water on the paintbrush pull together strongly because there are no molecules outside to pull on. So the bristles are pulled together too.

Page 28

The plasticine ball that falls the greatest distance will have the biggest dent. It has had more time to accelerate and will hit the ground at a greater speed than the others.

Page 31: Power puzzle

The work done by climbing stairs 10 metres high in 2 seconds, when you weigh 450 newtons = 450 × 10 = 4500 joules. The power used is 4500 ÷ 2 = 2250 watts.

Page 33: An electric question

When the negatively charged bottle is brought near to the plastic duck, the negative charges in the duck will be repelled. The negative charges move to the far end of the duck, leaving positive charges near the bottle. These will be attracted by the bottle, and the duck will follow the bottle. If the duck is negatively charged too, it will move away from the bottle.

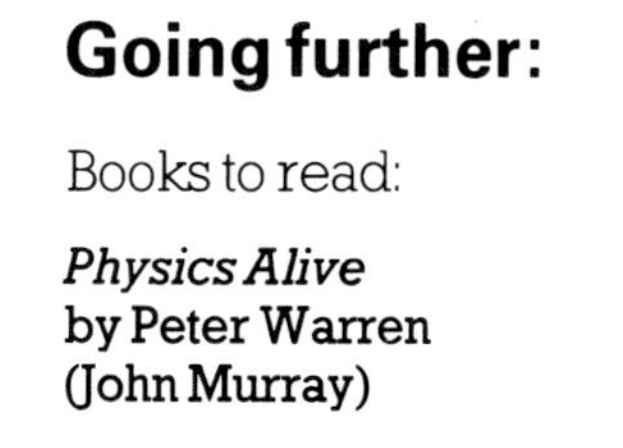

Going further:

Books to read:

Physics Alive
by Peter Warren
(John Murray)

Physics for You 1 & 2
by Keith Johnson
(Hutchinson)

The Young Scientist Book of Electricity
by Phil Chapman
(Usborne)

Physics for All
by J. J. Wellington
(ST(P))

Index